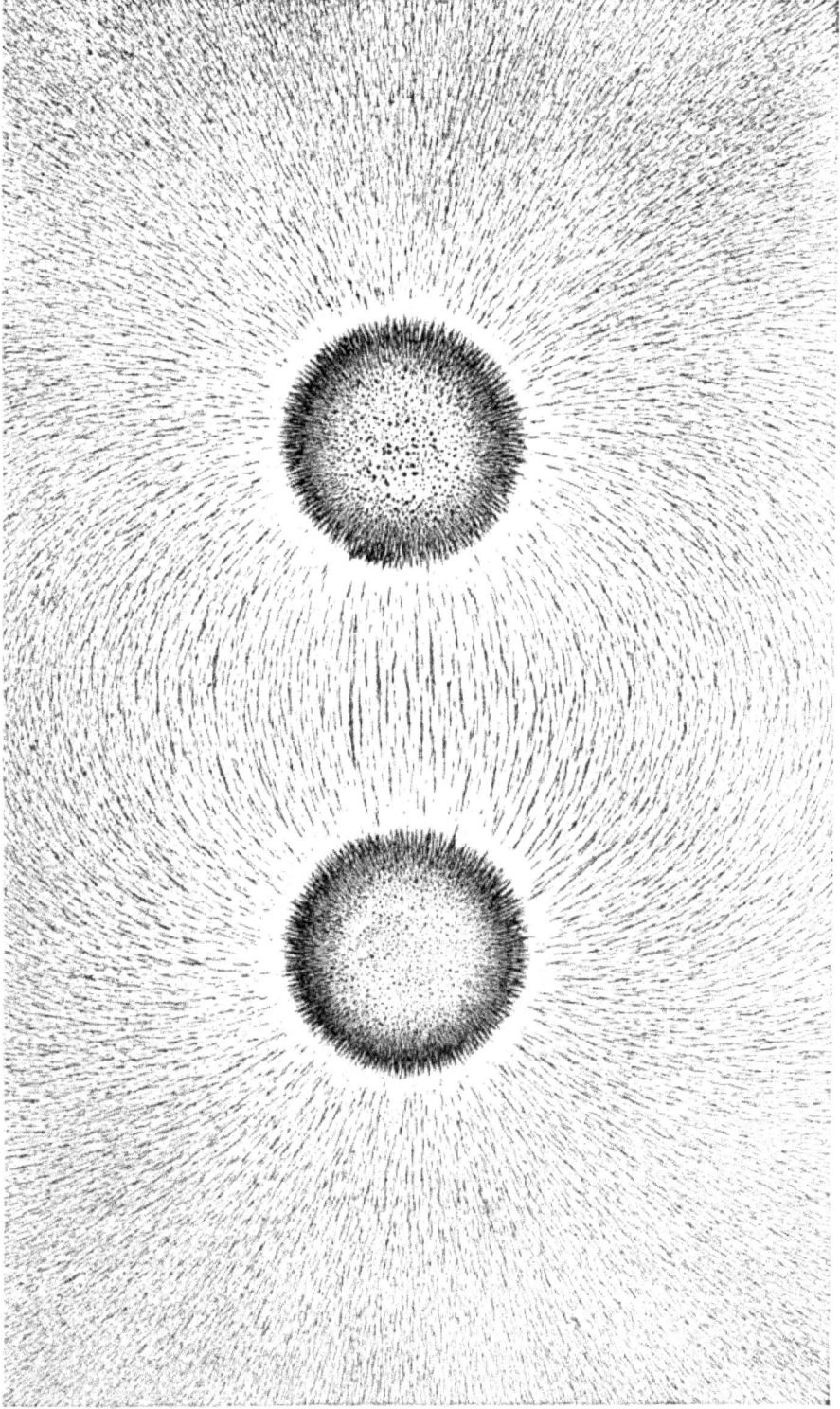

FRONTISPIECE.—The Magnetic Field. (One of the Schoonmaker Plates.)

SOLENOIDS
ELECTROMAGNETS

AND

ELECTROMAGNETIC WINDINGS

BY

CHARLES R. UNDERHILL

CHIEF ELECTRICAL ENGINEER, THE ACME WIRE COMPANY
FELLOW AMERICAN INSTITUTE OF ELECTRICAL ENGINEERS
MEMBER AMERICAN SOCIETY MECHANICAL ENGINEERS

224 ILLUSTRATIONS

SECOND EDITION, THOROUGHLY REVISED

THIRD THOUSAND

Merchant Books

1918

PREFACE TO FIRST EDITION

SINCE nearly all of the phenomena met with in electrical engineering in connection with the relations between electricity and magnetism are involved in the action of electromagnets, it is readily recognized that a careful study of this branch of design is necessary in order to predetermine any specific action.

With the rapid development of remote electrical control, and kindred electro-mechanical devices wherein the electromagnet is the basis of the system, the want of accurate data regarding the design of electromagnets has long been felt.

With a view to expanding the knowledge regarding the action of solenoids and electromagnets, the author made numerous tests covering a long period, by means of which data he has deduced laws, some of which have been published in the form of articles which appeared in the technical journals.

In this volume the author has endeavored to describe the evolution of the solenoid and various other types of electromagnets in as perfectly connected a manner as possible.

In view of the meager data hitherto obtainable it is believed that this book will be welcomed, not only by the electrical profession in general, but by the manufacturer of electrical apparatus as well.

The thanks of the author are due to Mr. W. D. Weaver, editor of *Electrical World,* for his permission to reprint articles, forming the basis of this work, originally published in that journal, and also for his friendly coöperation and encouragement. The labors of Professor Silvanus P. Thompson in this field deserve recognition from the electrical profession, to which the author desires to add his personal acknowledgments. The author's thanks are also due to the many friends to whose friendship he is indebted for the facilities afforded him to make the tests referred to in this volume. To Mr. Townsend Wolcott the author is indebted for his valuable assistance in correcting errors and for many suggestions.

CHARLES R. UNDERHILL.

NEW YORK,
   June, 1910.

PREFACE TO SECOND EDITION

THE extensive development in the control and mechanical utilization of electrical energy has given rise to the manufacture and supply in great quantities of electromagnets of multifarious forms, so that their design has become of great importance from an economical standpoint. The physical impossibility of insulating magnetic flux and the intricate mathematics which is required in the treatment of its spatial relations make it desirable to treat electromagnet problems in a somewhat empirical rather than in an involved theoretical manner. The present edition contains a full collection of data and makes suggestions as to the methods to be used in such a treatment. The data contained in this volume have been largely determined by experiments, there having been heretofore little practical published data on this subject. The experiments relating to the static pull are of particular value, since they give a concrete idea of the relationship between magnetomotive force and static pull. The so-called solenoid pull or leakage pull is of the greatest importance in the design of electromagnets with movable cores, and by means of these experimental data it is possible to very closely predetermine the mechanical forces due to such electromagnets. It is very clearly shown that the total mechanical pull or force is made up of two components which must be treated separately and finally united in order to find the total static pull through all points of the range or travel of the core. The mechanical forces or pulls due to alternating-current electromagnets generally follow the same laws as those for direct-current electromagnets, only that the current increases in proportion to the length of the air-gap, in the former, so that by taking the proper ratios between the direct and alternating currents, and allowing for iron losses, the average static pulls due to alternating-current electromagnets may also be closely approximated.

In this edition all errors have been corrected and additional wire tables given.

CHARLES R. UNDERHILL.

New Haven, Conn.,
    May 1, 1914.

Merchant Books

CONTENTS

CHAPTER I

INTRODUCTORY

CHAPTER II

MAGNETISM AND PERMANENT MAGNETS

CHAPTER III

ELECTRIC CIRCUIT

CHAPTER IV

ELECTROMAGNETIC CALCULATIONS

CHAPTER V

THE SOLENOID

CHAPTER VI

PRACTICAL SOLENOIDS

CHAPTER VII

IRON-CLAD SOLENOID

CHAPTER VIII

PLUNGER ELECTROMAGNETS

CHAPTER IX

ELECTROMAGNETS WITH EXTERNAL ARMATURES

CHAPTER X

ELECTROMAGNETIC PHENOMENA

CONTENTS
xi

CHAPTER XVIII

ELECTROMAGNETIC WINDINGS

CHAPTER XIX

FORMS OF WINDINGS AND SPECIAL TYPES

CHAPTER XX

HEATING OF ELECTROMAGNETIC WINDINGS

CHAPTER XXI

TABLES AND CHARTS

LIST OF ILLUSTRATIONS

MB

SOLENOIDS, ELECTROMAGNETS, AND ELECTROMAGNETIC WINDINGS

CHAPTER I

INTRODUCTORY

1. DEFINITIONS

Force is that which produces or tends to produce motion.

Resistance is whatever opposes the action of a force.

Work is the overcoming of resistance continually occurring along the path of motion.

Energy is the capacity for doing work; therefore, the amount of work that may be done depends upon the amount of energy expended.

The *Effective Work* is the actual work accomplished after overcoming friction.

Time is the measure of duration.

Power is the rate of doing work, and is equal to work divided by time.

It is to be noted that work does not embrace the time factor; that is, no matter whether a certain amount of work requires one minute or one month to accomplish, the value of work will be the same.

With power, however, time is an important factor; for, if a certain amount of work is to be accomplished by one machine in one half the time required by another, the former will require twice the power required in the latter.

The product of power into time equals the amount of work.

Efficiency is the ratio between the effective work and the total energy expended. It is usually expressed as a percentage.

2. The C. G. S. System of Units

The *Centimeter-Gram-Second* system embraces the *Centimeter* as the unit of length, the *Gram* as the unit of mass, and the *Second* as the unit of time. These are the *Fundamental* units.

The centimeter is 0.01 *Meter*, the meter being $\frac{1}{10,000,000}$ part of the earth-quadrant through the meridian of Paris, measured from the Equator to the North Pole. The equivalent of the meter is, in English measure, 39.37 inches. Therefore, 1 centimeter = 0.3937 inch.

The *Gram* is equal to one cubic centimeter of distilled water at its maximum density, which is at 4° Centigrade. Mass is a constant, but weight varies at different places according to the force of gravitation at those places. The equivalent of the gram in English measure is 0.00220464 pound.

The *Second* is the $\frac{1}{86,400}$ part of the mean solar day.

The *Absolute* units are based upon the fundamental units.

The *Dyne* is the absolute unit of force, and is that force which, acting upon one gram for one second, imparts to it a velocity of one centimeter per second. The *pull* due to gravity on 1 gram = 981 dynes.

The *Erg* is the absolute unit of work, and is the work done when one dyne acts through one centimeter.

The following prefixes are used in the C. G. S. system.

Mili meaning thousandth part.

Centi meaning hundredth part.

Deci meaning tenth part.

Deca meaning ten.

Hecto meaning one hundred.

Kilo meaning one thousand.

Thus the centimeter is the one hundredth part of the meter; the kilometer is one thousand meters, etc.

Abbreviations for the metric units are m. for meter, cm. for centimeter, mm. for milimeter, g. for gram, kg. for kilogram, etc.

3. GENERAL RELATIONS BETWEEN COMMON SYSTEMS OF UNITS

In the English system of units the mechanical unit of work is the *Foot-pound*, and is the amount of work required to raise one pound vertically one foot.

The mechanical unit of power is the *Horse-power*, and is the power required to raise 33,000 pounds one foot vertically, in one minute, or, in other words, 33,000 foot-pounds per minute.

Since the laws of electrical engineering are expressed in terms of the C. G. S. units, these units should be used as much as possible in all calculations.

Figures 1 to 3 show the relations between the English and C. G. S. units most commonly used.

In general it may be stated that the calculations of the magnetic circuit may be made in metric units, while

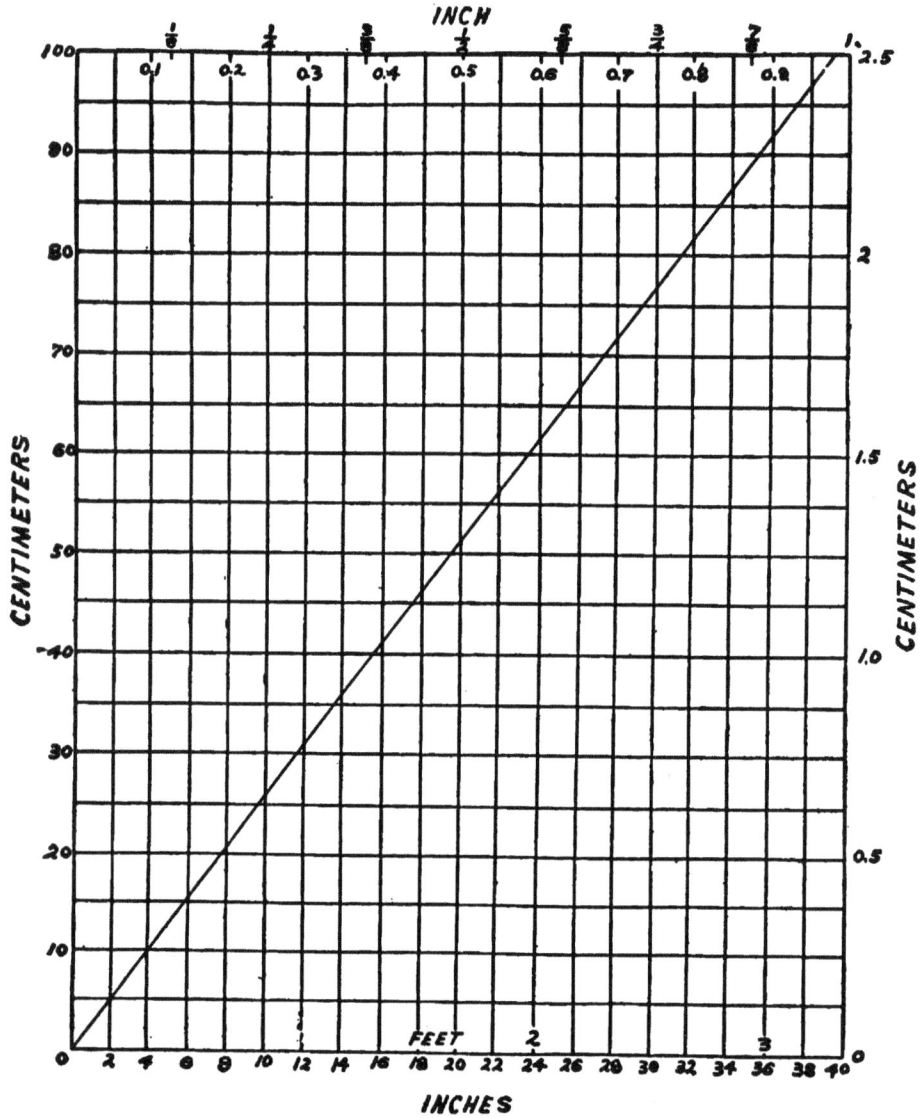

FIG. 1. — Conversion Chart. Linear.

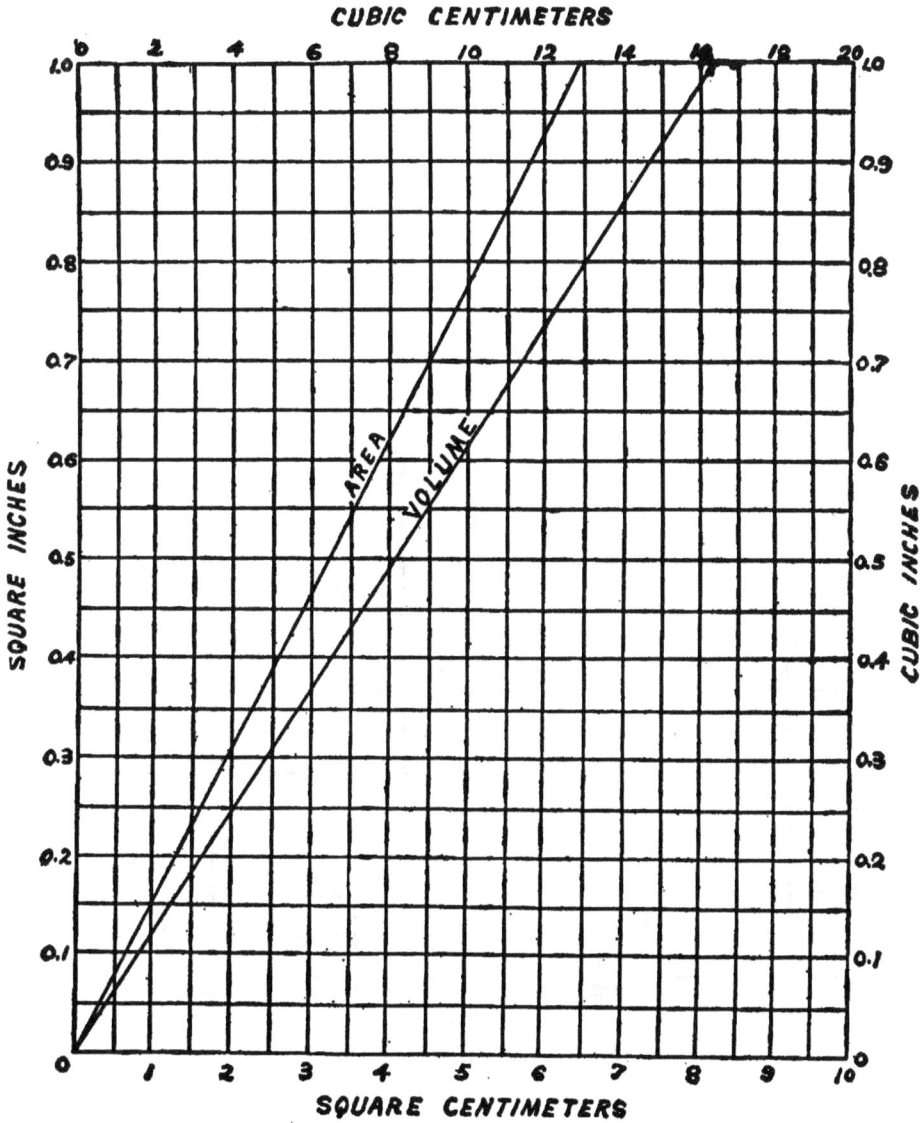

FIG. 2. — Conversion Chart. Area and Volume.

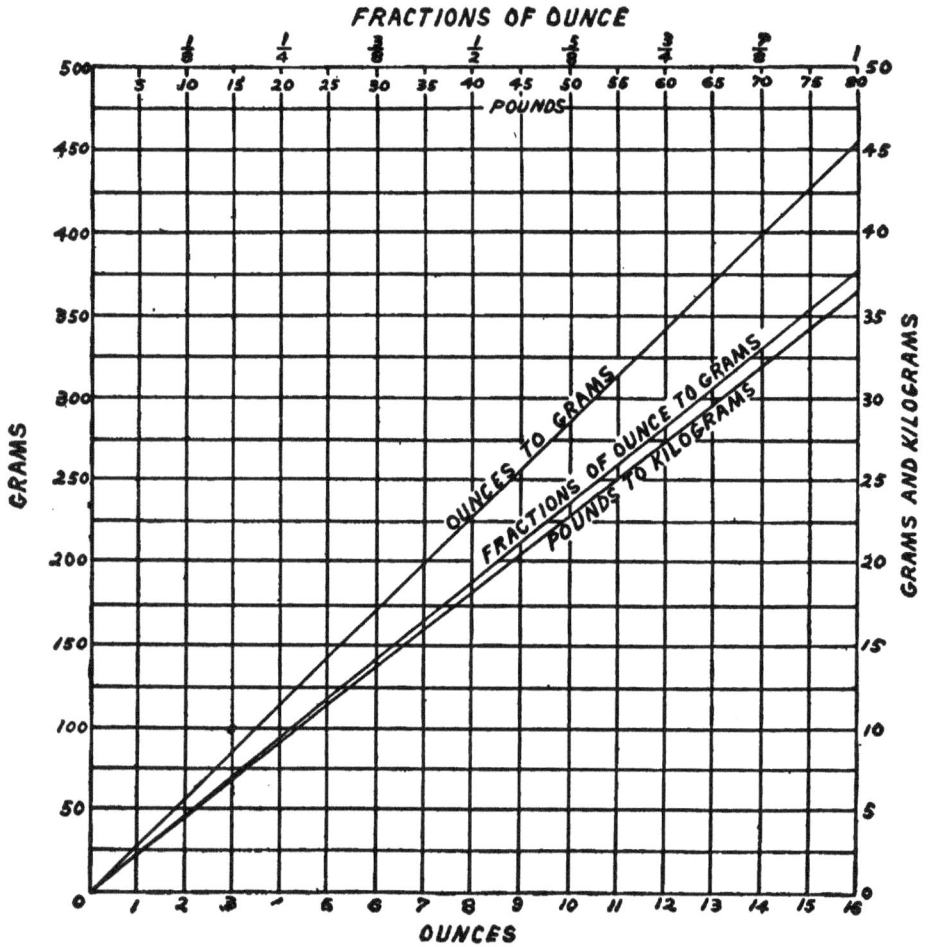

FIG. 3. — Conversion Chart. Weights.

it may be more convenient to express the dimensions of the winding, diameter of wire and thickness of insulation by English units, since nearly all obtainable data for insulated wires are given in the latter units. However, the formulæ in this book are so arranged that either system may be used. By the use of the charts in Figs. 1 to 3 conversions may be easily made.

4. Notation in Powers of Ten

Instead of writing a number like ten millions thus: 10,000,000, it is often more convenient to express it thus: 10^7. Therefore, the number 2,140,000 may be written 214×10^4, or 2.14×10^6.

Likewise, $\frac{1}{10}$ may be expressed 10^{-1}, and $\frac{1}{1,000,000}$, 10^{-6}, etc.

CHAPTER II

MAGNETISM AND PERMANENT MAGNETS

5. MAGNETISM

"*Magnetism* is that peculiar property occasionally possessed by certain bodies (more especially iron and steel) whereby under certain circumstances they naturally attract or repel one another according to determinate laws."

The ancients in Magnesia, Thessaly, are supposed to have been the original discoverers of magnetism, where an ore possessing a remarkable tractive power for iron was found. To a piece of this iron-attracting ore was given the name *Magnet*.

It was further found that a piece of this ore, when freely suspended, swung into such a position that its ends pointed north and south, which discovery made it possible for navigators to steer their ships by means of the *Lodestone* (leading stone).

A piece of hardened steel was found to possess the properties of the lodestone when the former was rubbed by the latter; thus becoming an *Artificial Magnet*.

There is no known insulator of magnetism; nearly all substances have the same conducting power as air, which, however, is not a very good conductor. A magnetic substance is one which offers little resistance to the *Magnetic Force;* that is, it is a good conductor of magnetism as compared with air. The conducting power of the all-pervading *Ether* is taken as unity, and is approximately the same as that of air.

6. Magnetic Field

Theory indicates, and experiment confirms, that magnetism flows along certain lines called *Lines of Force*, and that these always form closed paths or circuits. The region about the magnet through which these lines pass, is called the *Field of Force*, and the path through which they flow is called the *Magnetic Circuit*.

A magnet in the form of a closed ring (Fig. 4) will not attract other magnetic substances to it, since an excellent closed circuit or path is provided in the ring through which the lines of force pass. However, if this ring be separated, as in Fig. 5, the magnetic effect will be pronounced at the points of separation. The opposite halves of the ring will be strongly attracted, and magnetic substances, such as iron or steel, will be drawn to, and firmly held at, the points of separation.

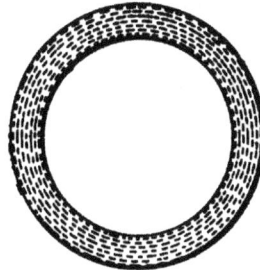

FIG. 4.

Closed Ring Magnet.

The reason for this is that when the magnetic ring is divided, a good path for the lines of force is no longer provided at these points; but, as the air possesses unit-conducting power, the lines pass through it and into the magnetic ring again.

When a magnetic substance is brought near the points of separation, however, this magnetic substance offers a better path for the lines of force than the air;

hence, as the magnetic field always tends to shorten itself, thus producing a stress, the magnetic substance will be drawn to the point of separation in the magnetized ring, and into such a position as to form the best conducting bridge across the *Air-gap*.

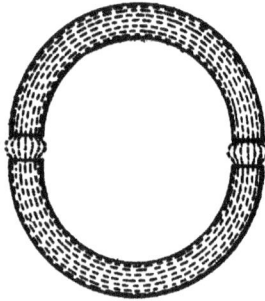

FIG. 5.

Separated Ring Magnet.

Quite a different effect is produced when the magnet is in the form of a straight bar. In this case only a part of the magnetic circuit consists of a magnetic substance; hence, the lines of force will pass out through the surrounding air before they can again enter the magnet.

The paths of the lines of force can be demonstrated by placing a piece of paper over a bar magnet and then sprinkling iron filings over the paper, which should be jarred slightly in order that the filings may be drawn into the magnetic paths. This effect is shown in Fig. 6.*

7. PERMANENT MAGNETS

Artificial magnets which retain their magnetism for a long time are called *Permanent Magnets*. These are made by magnetizing hardened steel, the hardening process tending to cause the molecules of the steel to permanently remain in one direction when magnetized. It is assumed that in soft iron or steel the molecules normally lie in such positions as to neutralize any magnetic tendency on the part of the material as a whole.

* Made for this volume by Mr. E. T. Schoonmaker.

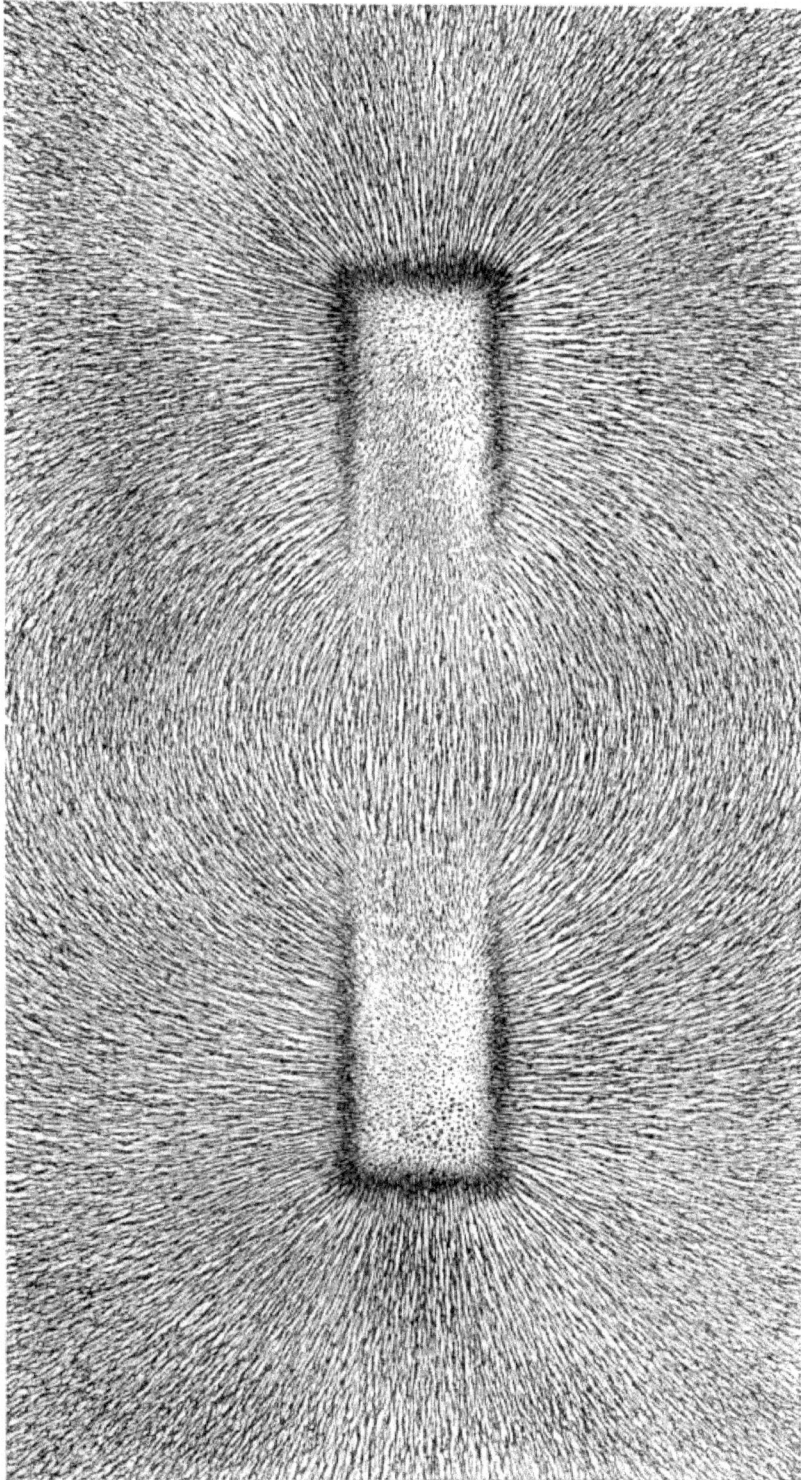

Fig. 6. — Field of Force surrounding Magnet.

When the soft iron or steel is placed in a sufficiently strong magnetic field, the molecules readily lie end to end, so to speak; thus possessing all the properties necessary for a magnet. However, the molecules assume (approximately) their normal positions as soon as the magnetizing influence is removed; hence, the steel must be hardened to produce a good permanent magnet.

Permanent magnets are used in electrical testing instruments where a constant magnetic field is required, and also as the *Field Magnets* of magnetos, such as are extensively used on automobiles and in telephone apparatus. As these magnets have a tendency to deteriorate with age, they are artificially *aged* by placing them in boiling water for several hours.

When an externally applied magnetizing force is withdrawn, the magnetism does not wholly disappear, and that force which tends to retain the magnetism is known as *Retentiveness*, while that portion of magnetization remaining is called *Residual Magnetism*. The magnetizing force necessary to remove all residual magnetism is called the *Coercive Force*. Soft iron has little coercive force, but great retentiveness; while hardened steel has great coercive force as well as retentiveness.

8. Magnetic Poles

Although the term *North Pole* is given to that end of a bar magnet which points north, we will, in this book, make use of the term *North-seeking Pole* instead of the former term in order to avoid confusion between the north pole of a magnet and the pole situated near the North Pole of the earth.

The strengths of the north-seeking and south-seeking poles of a magnet are equal; the strength diminishing gradually from the ends to the center or *Neutral Point* of the magnet, where there is no attraction whatever.

Unlike poles attract, while like poles repel one another.

Magnetism flows from the north-seeking pole of a magnet, through the surrounding region to its south-seeking pole, and thence through the inside of the magnet to the north-seeking pole. Reference to Fig. 6 shows that all of the magnetic lines do not flow from the ends of the magnet, but from all points on the north-seeking portion to corresponding points on the south-seeking portion.

The theoretical pole of a magnet is regarded as a point and not as a surface; hence, in practice the term *Pole* is better applied to the surface where the density of the lines entering or leaving the magnet is greatest. The direction in which the lines of force flow indicates the *Polarity* of the magnet as previously described. This explains why every magnet has two poles. It is evident, then, that no matter into how many pieces a permanent magnet may be separated, each piece will be a magnet, since the coercive force remains in each piece and the lines leave at one part and enter at another. Both poles will, therefore, be of equal strength.

4π lines of force radiate from a unit magnetic pole; for, if this pole be placed at the center of a sphere of one centimeter radius, one line of force per square centimeter will radiate from this pole, and the area of the sphere is $4\pi r^2$ square centimeters.

9. FORMS OF PERMANENT MAGNETS

What may be called the natural form of permanent magnet is shown in Fig. 7. This is known as the *Bar Permanent Magnet*, and is the form which constitutes

FIG. 7.—Bar Permanent Magnet.

the compass needle. It is not, however, an efficient form for most purposes, owing to the fact that its effective polar regions are widely separated.

The practical permanent magnet consists of a bar magnet bent into the form of U, so as to shorten the magnetic circuit by bringing the polar regions of the magnet close together. This is called a *Horseshoe* permanent magnet, and is shown in Fig. 8.

A permanent magnet does work when it attracts a piece of iron or other magnetic substance, called its *Armature*, to it.

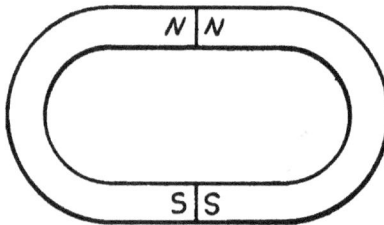

FIG. 8.

Horseshoe Permanent Magnet.

When the armature is forcibly removed from the magnet, however, energy is returned to the magnet. Since the effective strength of a magnet varies inversely as the resistance to the magnetic force, the air-gaps should be as small as possible. This is equivalent to stating that there is greater attraction between a magnet and its armature through a short than through a greater distance.

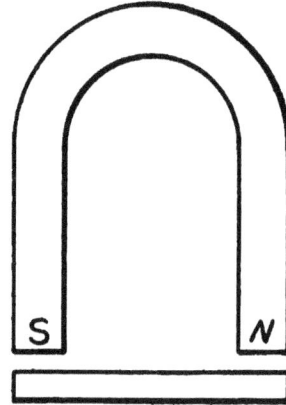

FIG. 9.

Magnet with Consequent Poles.

Another type of horseshoe magnet is shown in Fig. 9. This is said to have *Consequent Poles*, since the ends of similar polarity are placed together. The

same effect may be obtained
with the arrangement in Fig.
10. It is important, however,
that the individual magnets
constituting the *Compound
Magnet* should have the same
strength in order that one

FIG. 10.

Compound Magnet.

magnet may not act as a return circuit for the other,
thus weakening the combination.

10. MAGNETIC INDUCTION

When a piece of iron is attracted by a magnet, it also
temporarily becomes a magnet, and a series of pieces of
iron will attract one another successively so long as
the first piece is influenced by the magnet. This phenomenon is said to be the result of *Magnetic Induction*.
In this case the pieces of iron tend to form a good conducting path for the lines of force; hence, the more
perfectly they tend to close the magnetic circuit, the
greater will be their attraction for one another.

11. MAGNETIC UNITS

Unit Strength of Pole is that which repels another similar and equal pole with unit force (one dyne) when placed
at a unit distance (one centimeter) from it. (Symbol m.)

Magnetic Moment (symbol $\mathfrak{M}$) is the product of the
strength of either pole into the distance between the poles.

Intensity of Magnetization (symbol $\mathfrak{I}$) is the magnetic moment of a magnet divided by its volume.

$$\mathfrak{M} = lm, \quad (1) \qquad\qquad \mathfrak{I} = \frac{\mathfrak{M}}{v}, \qquad (2)$$

wherein l = distance between poles
and v = volume of magnet.

Intensity of Magnetic Field (symbol $\mathcal{H}$) is measured by the force it exerts upon a unit magnetic pole, and, therefore, the unit is the intensity of field which acts upon a unit pole with unit force (one dyne). The unit is the *Gauss*. Hence, one gauss is one line of force per square centimeter.

$$\mathcal{H} = \frac{\mathcal{C}}{\mathcal{M}}. \tag{3}$$

Magnetic Flux (symbol ϕ) is equal to the average field intensity multiplied by the area. Its unit (one line of force) is the *Maxwell*.

One gauss is, therefore, equal to one maxwell per square centimeter.

Reluctance or Magnetic Resistance (symbol $\mathcal{R}$) is the resistance offered to the magnetic flux by the material magnetized. The unit is the *Oersted*, and is the reluctance offered by a centimeter cube of vacuum.

Magnetic Induction or Flux Density (symbol $\mathcal{B}$) is the number of magnetic lines per unit area of cross-section of magnetized material, the area being at every point perpendicular to the direction of flux. The unit is the gauss.

Magnetic Permeability (symbol μ) is the ratio of the magnetic induction $\mathcal{B}$ to the field intensity $\mathcal{H}$, and is the reciprocal of *Reluctivity* (specific magnetic reluctance).

CHAPTER III

ELECTRIC CIRCUIT

12. UNITS

Resistance (symbol R) is that property of a material that opposes the flow of a current of electricity through it. The practical unit is the *Ohm*, and its value in C. G. S. units is 10^9.

Electromotive Force (e. m. f., symbol E) is the electric pressure which forces the current through a resistance. The unit is the *Volt*, and its value in C. G. S. units is 10^8.

Difference of Potential is simply a difference of electric pressure between two points. The unit is the *Volt*.

Current (symbol I) is the intensity of the electric current that flows through a circuit. The unit is the *Ampere*. Its value in C. G. S. units is 10^{-1}.

Ohm's Law. The strength of the current is equal to the electromotive force divided by the resistance, or

$$I = \frac{E}{R} \quad (4), \text{ whence } R = \frac{E}{I} \quad (5), \text{ and } E = IR. \quad (6)$$

Conductance (symbol G) is the reciprocal of resistance. The practical unit is the *Mho*.

Since $G = \dfrac{1}{R}$ (7), $I = EG$ (8), $G = \dfrac{I}{E}$ (9), and $E = \dfrac{1}{G}$ (10).

Electric Energy (symbol W) is represented by the work done in a circuit or conductor by a current flowing through it. The unit is the *Joule*, its absolute value is 10^7 ergs, and it represents the work done by the flow, for one second, of 1 ampere through 1 ohm.

Electric Power (symbol P) is 1 joule per second. The unit is the *Watt* and equals 10^7 absolute units. 745.6 watts equal 1 horse-power. 1 *Kilowatt* equals 1000 watts.

Hence, $\quad\quad\quad W = PT,$ $\quad\quad\quad\quad\quad\quad\quad\quad$ (11)

wherein $\quad\quad\quad\quad W =$ energy in joules,

$\quad\quad\quad\quad\quad\quad\quad P =$ power in watts,

and $\quad\quad\quad\quad\quad T =$ time in seconds.

Now $P = EI$ (12). Substituting the value of P from (12) in (11), $W = EIT$. $\quad\quad\quad\quad\quad$ (13)

$$1 \text{ watt} = \begin{cases} 6.119 \text{ kilogrammeters per minute,} \\ 44.26 \text{ foot-pounds per minute,} \\ 0.001 \text{ kilowatt,} \\ 0.00134 \text{ horse-power.} \end{cases}$$

Also $P = I^2R$ (14), $= \dfrac{E^2}{R}$. $\quad\quad\quad\quad\quad\quad$ (15)

Density of Current in a conductor is equal to the total current in amperes divided by the cross-sectional area of the conductor, or $I_d = \dfrac{I}{A_w}$. $\quad\quad\quad\quad$ (16)

When a current of electricity flows through a conductor, heat is generated, due to electrical friction in the conductor, and is directly proportional to the watts lost in the conductor. The resistance of a conductor changes with its temperature. In nearly all cases the resistance rises with the temperature. The ratio between rise in temperature and rise in *Resistivity* (specific resistance) is called the *Temperature Coefficient*, which for copper is approximately 0.00388 at 20° C. or 68° F.; that is, the resistance will change approximately 0.388 per cent for each degree Centigrade change in temperature. The chart on p. 300 gives the coefficients for different temperatures.

Referring to equation (6), it is evident that the voltage or e. m. f. per unit length of conductor is proportional to the resistance per unit length, and the strength of the current flowing through the conductor.

The resistance is equal to the length of the conductor divided by its cross-sectional area and *Conductivity* (specific conductance), or

$$R = \frac{l_w}{A_w \gamma}, \qquad (17)$$

wherein l_w = length of conductor,

$\qquad A_w$ = cross-sectional area of conductor,

and $\qquad \gamma$ = conductivity of material.

Substituting the value of R from (17) in (4),

$$I = \frac{E}{\left(\frac{l_w}{A_w \gamma}\right)}, \qquad (18)$$

and (6) then becomes $\qquad E = I \frac{l_w}{A_w \gamma}. \qquad (19)$

Where the temperature is subject to considerable change due to either internal or external influences, the value of γ will vary; hence, equation (17) must be used.

It is also evident that where two or more conductors of different conductivities, and particularly if of the same cross section, form part of the same electric circuit, the greater part of the e. m. f. will be expended in overcoming the resistance of the conductor having the lowest value for γ. It will also be evident that for constant e. m. f., the conductor having the greater conductivity will have the greater density of current when singly connected.

When a conductor containing resistance is connected with a source of electrical energy, such as a battery, which also offers some resistance, the conductor will receive the maximum amount of electrical energy when its resistance equals the sum of all the other resistances in the circuit. The rule is somewhat modified under certain conditions, as will be seen by referring to p. 287.

13. CIRCUITS

When two or more conductors are connected as in Fig. 11, they are said to be in *Series*, and in *Multiple* when connected as in Fig. 12. The latter is called a *Shunt* circuit.

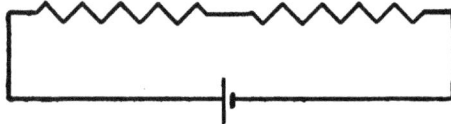

FIG. 11. — Resistances in Series.

Consider two conductors, each having a resistance of 100 ohms. When connected in series, the total resistance would be $2 \times 100 = 200$ ohms. If connected in multiple, the *Joint Resistance* would be $\frac{100}{2} = 50$ ohms.

Hence, the series resistance is four times as great as the multiple resistance.

When any number of resistances are connected in series, the total

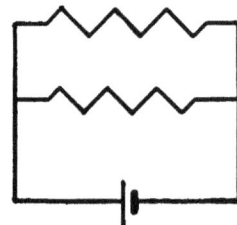

FIG. 12.

Resistances in Multiple.

resistance will be the sum of the individual resistances. When any number of *equal* resistances are connected in multiple, their joint resistance will be the common resistance of all the circuits divided by the number of circuits.

When two equal or unequal resistances are connected in multiple, their joint resistance is equal to their product divided by their sum, or

$$R_j = \frac{RR_1}{R + R_1}.$$ (20)

When any number of equal or unequal resistances are connected in multiple, the joint resistance is equal to the reciprocal of their joint conductance. Consider three resistances R_1, R_2, and R_3. The conductances are

$$\frac{1}{R_1}, \frac{1}{R_2}, \text{ and } \frac{1}{R_3}.$$

Their *Joint Conductance* is

$$G_j = \frac{1}{R_1} + \frac{1}{R_2} + \frac{1}{R_3} = \frac{R_2R_3 + R_1R_3 + R_1R_2}{R_1R_2R_3}.$$ (21)

Now $R_j = \dfrac{1}{G_j}.$ (22)

Hence,

$$R_j = \frac{R_1R_2R_3}{R_2R_3 + R_1R_2 + R_1R_3}.$$ (23)

In Fig. 13 is shown a circuit consisting of a source of electrical energy with an internal resistance of 2 ohms, an external resistance of 2.723 ohms, and a shunt

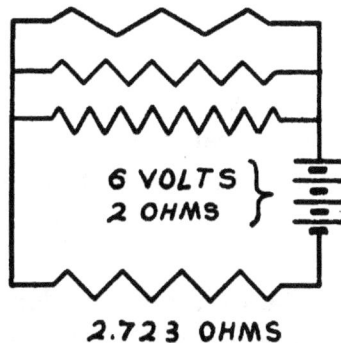

6 VOLTS
2 OHMS

2.723 OHMS

FIG. 13. — Divided Circuit in Series with Resistance.

circuit consisting of three resistances of 3, 4, and 5 ohms respectively.

The combination is connected in series, forming a circuit partly in series and partly in multiple. The e. m. f. of the battery is 6 volts.

The joint resistance of the shunt circuit is, from (23),

$$R_j = \frac{3 \times 4 \times 5}{4 \times 5 + 3 \times 4 + 3 \times 5} = \frac{60}{47} = 1.277 \text{ ohms.}$$

The total resistance in series is, therefore,

$$2 + 2.723 + 1.277 = 6 \text{ ohms.}$$

Since $E = 6$, $\qquad\qquad I = \frac{6}{6} = 1$ ampere.

Since $E = IR$, the difference of potential (*drop* in volts) across each resistance will be as follows:

Drop in battery	$= 1 \times 2$	$= 2$ volts
Drop in series resistance	$= 1 \times 2.723$	$= 2.723$ volts
Drop in shunt circuit	$= 1 \times 1.277$	$= 1.277$ volts
Total drop		$= 6.000$ volts

It is thus seen that the resistances in the series circuit can be considered as *counter e. m. f.'s*, and added together.

In shunt circuits all conductances may be considered as currents and added together. Hence, in the above case, the conductances are

$$\tfrac{1}{3} = 0.333, \ \tfrac{1}{4} = 0.25, \text{ and } \tfrac{1}{5} = 0.20.$$

The total current flowing through all the branches at a pressure of 1 volt would, therefore, be

$$I = 0.333 + 0.25 + 0.20 = 0.783 \text{ ampere.}$$

However, in the case considered, the e. m. f. is 1.277 volts. Hence, the total current will be $0.783 \times 1.277 = 1$ ampere, and the actual current flowing through each branch will be

$$1.277 \times 0.333 = 0.425 \text{ ampere,}$$
$$1.277 \times 0.250 = 0.318 \text{ ampere,}$$
$$1.277 \times 0.200 = 0.257 \text{ ampere,}$$

or a total of 1 ampere.

CHAPTER IV

ELECTROMAGNETIC CALCULATIONS

14. ELECTROMAGNETISM

WHEN a compass needle is placed near a conductor through which an electric current is flowing, the needle tends to assume a position at right angles to the current in the wire. If the needle is above the wire, and the current flows from left to right, the north-seeking pole is deflected toward the observer. If the needle is below the wire, the north-seeking pole is deflected from the observer.

15. FORCE SURROUNDING CURRENT IN A WIRE

When an electric current flows, it establishes a magnetic field at right angles to it in the form of concentric

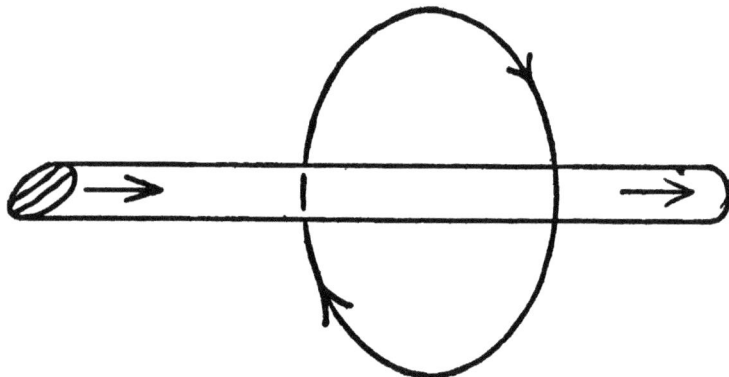

FIG. 14.—Relation between Directions of Current and Force surrounding It.

circles of force. The compass needle, being a magnet, is drawn by mutual attraction into such a position that its magnetic circuit will lie in the same direction as the lines of force about the current. The earth's magnetism, unless neutralized, tends to prevent the needle from lying exactly in the direction of the lines of force due to the current.

The relation between the directions of current and flux are shown in Fig. 14.

16. ATTRACTION AND REPULSION

Two wires lying parallel to one another, and carrying currents in the same direction, will be mutually attracted; while if the currents are opposite in direction, they will be mutually repelled.

If a conductor carrying a current be placed between the poles of a horseshoe magnet, and at right angles to the lines of force, it will be either attracted or repelled, according to the relative directions of the field due to the magnet and that due to the current in the wire.

By placing a loop of wire, through which a current is flowing, around a bar permanent magnet the same general action will result. If the polarities of the loop and magnet are the same, the loop will tend to remain at the center of the magnet; whereas, if the polarities are opposite, the loop will be repelled from the magnet and will not remain in any position around it.

The relation between the strength of current in a wire and the intensity of magnetic field or *Magnetizing Force* is expressed by the equation

$$\mathscr{H} = \frac{0.2\,I}{a}, \qquad (24)$$

wherein

$\mathcal{H} =$ magnetizing force in gausses,

$I =$ current in amperes,

and $a =$ radial distance from center of wire in centimeters.

17. FORCE DUE TO CURRENT IN A CIRCLE OF WIRE

Under these conditions the lines of force are distorted, as in Fig. 15, but in the center

$$\mathcal{H} = \frac{0.2\,\pi I}{r}, \quad (25)$$

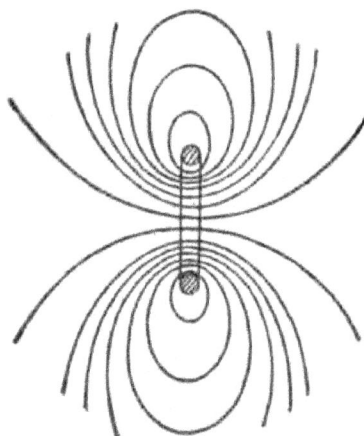

FIG. 15. — Distortion of Field due to Circular Current.

wherein r is the average radius of the turn of wire, as in Fig. 16.

At any distance x on the axis X the force is

$$\mathcal{H} = \frac{0.2\,\pi I r^2}{S^3}. \quad (26)$$

This will be better understood if we consider the force at the center by this formula. At the center $S = r$. Therefore, substituting r for S, and referring to equation (26),

$$\mathcal{H} = \frac{0.2\,\pi I r^2}{r^3}, \quad (27) \qquad = \frac{0.2\,\pi I}{r}. \quad (25)$$

Since
$$S = \sqrt{r^2 + x^2}, \quad (28)$$

$$\mathcal{H} = \frac{0.2\,\pi I r^2}{(r^2 + x^2)^{\frac{3}{2}}}. \quad (29)$$

From (25) is deduced the following law: *If a wire one centimeter in length be bent into an arc of one centi-*

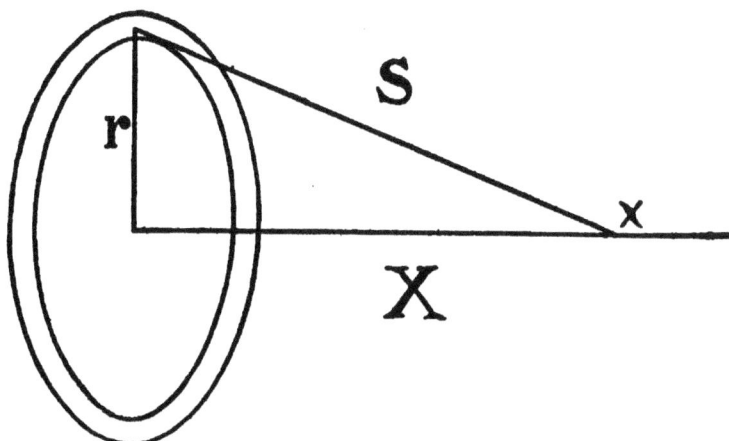

FIG. 16. — Strength of Field at Varying Distances from Center of Loop.

meter radius, and a current of 10 amperes passed through the wire, at the center of the arc there will be one line of force per square centimeter, i.e. *the intensity will be one gauss.*

18. AMPERE-TURNS

One ampere flowing through one turn of wire is called one *Ampere-turn.* In any case the product of I amperes into N turns of wire equals ampere-turns. Hence, the symbol is IN. Since the absolute unit of current is equivalent to 10 amperes, this current flowing through one turn of wire equals 10 ampere-turns, and this produces 4π dynes, which is the total force due to the magnetic pole of unit strength.

19. THE ELECTROMAGNET

Electric and magnetic circuits differ in the respect that there is no known insulator of magnetism, while electricity can be insulated. Thus, while dry air may effectually insulate electricity, it serves as a unit-conducting medium for magnetism.

Magnetism cannot be efficiently transmitted over any great distance on account of leakage. The practical method is to transmit a current of electricity through a wire, and then convert its energy into magnetism at the point where the attraction is desired.

This is accomplished by winding spirals of insulated wire around the magnetic material which is to be magnetized. Such a device is known as an *Electromagnet*, and upon the passage of an electric current through the winding, the magnetic material behaves similarly to a permanent magnet of the same general form, with the exception that, if the magnetic material has but little coercive force, the magnetism will practically disappear upon the discontinuation of the electric current through the winding.

20. EFFECT OF PERMEABILITY

The permeability or magnetic conductivity of magnetic materials, such as iron or steel, decreases as the flux density increases. The relation is expressed

$$\mu = \frac{\mathcal{B}}{\mathcal{H}} \tag{30}$$

wherein μ = permeability,

$\mathcal{B}$ = magnetic induction or flux density,

and $\mathcal{H}$ = magnetizing force or intensity of field.

Figure 17 shows the variation in permeability for different values of $\mathcal{B}$. This is called the *Permeability Curve*. Such curves are obtained from iron and steel by actual tests, and these data used in subsequent calculations. The table on p. 323 will be found useful.

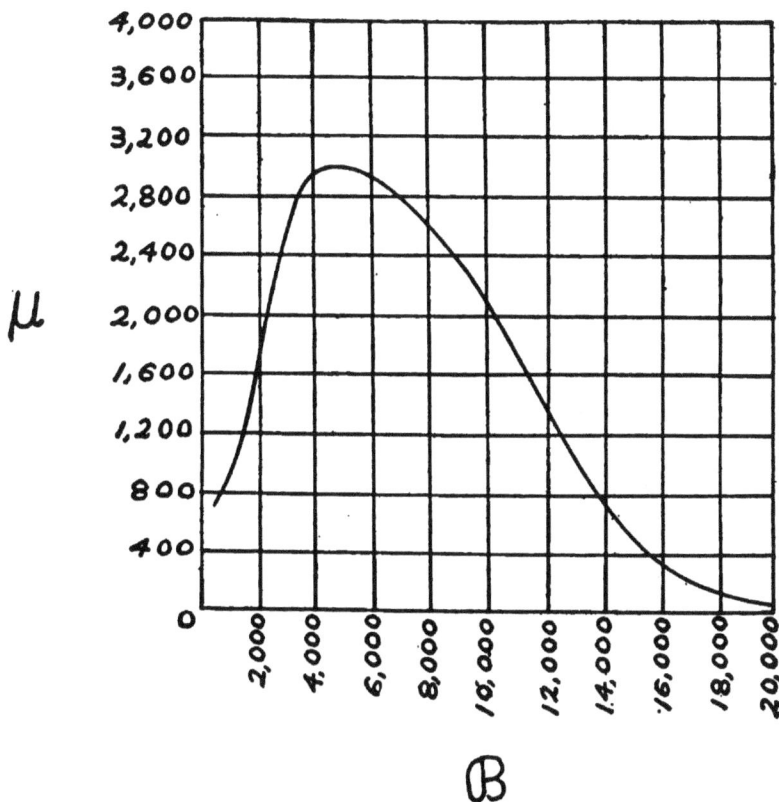

FIG. 17. — Permeability Curve.

21. SATURATION

In Fig. 18 is shown the general relations between $\mathcal{H}$ and $\mathcal{B}$. This is known as the *Magnetization Curve*. Similar curves are shown in Fig. 20, which will be referred to later. The point where the flux density or induction $\mathcal{B}$ is not materially increased by a considerable increase in the magnetizing force $\mathcal{H}$ is called the *Saturation Point*, or *Limit of Magnetization*, and at this point the iron or steel is said to be *Saturated*.

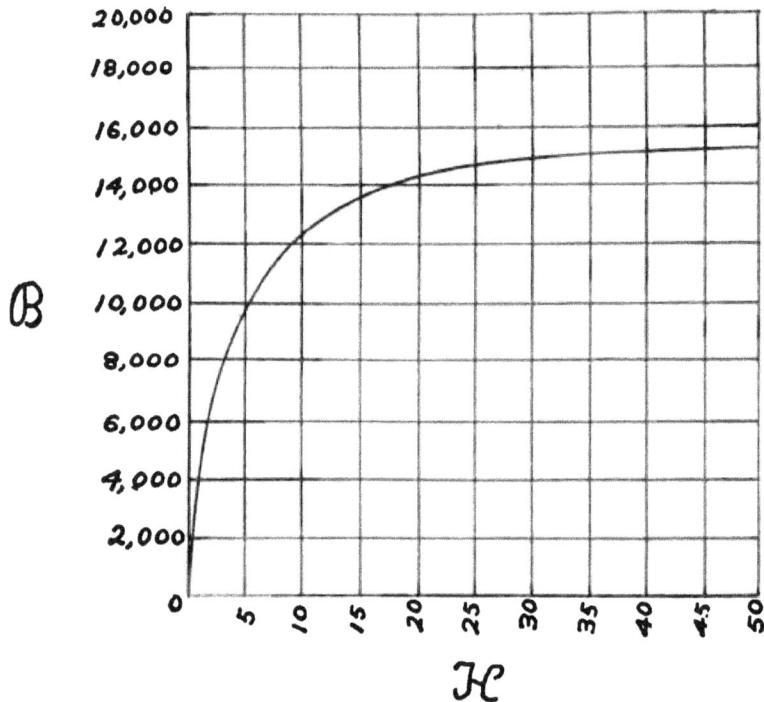

FIG. 18. — Magnetization Curve.

The values* of $\mathcal{B}$ at the saturation point, for various grades of iron and steel, are as follows:

Wrought iron 20,200
Cast steel 19,800
Mitis iron 19,000
Ordinary cast iron 12,000

The practical working densities are about two thirds of the above values and are as follows:

Wrought iron 13,500
Cast steel 13,200

* Wiener, *Dynamo-electric Machines.*

Mitis iron 12,700

Ordinary cast iron 8,000

The permeability should not fall below 200 to 300.

22. SATURATION EXPRESSED IN PER CENT

The amount of saturation of a given flux path may be conveniently expressed in per cent. It is obvious

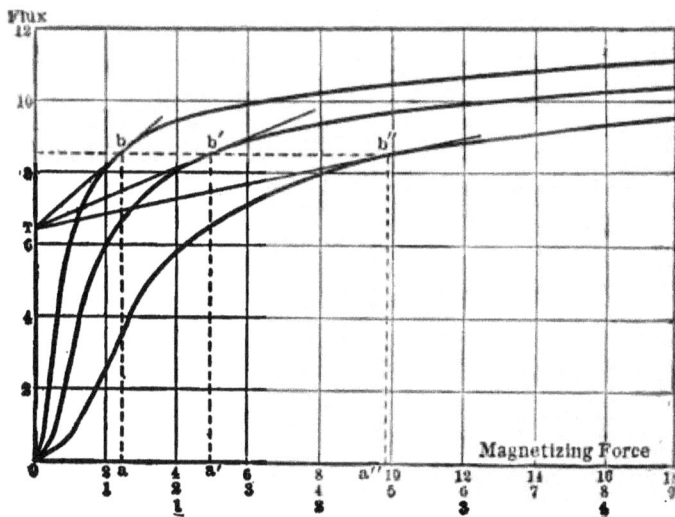

FIG. 19. — Saturation Curve plotted to Different Horizontal Scales.

that the statement that a magnet "is worked well up to the knee" or "well below the knee" means very little when any degree of accuracy is desired. This arbitrary method of defining saturation is deceptive because the position of "the knee" depends upon the scales to which the saturation curve is plotted, as will be seen by reference to Fig. 19, all the curves being plotted from the same data except that they are drawn to different scales.

What is desired is a definition which will indicate 100 per cent saturation for the condition in which there is no increase of flux for an increase of magnetizing force, and will indicate zéro saturation for the condition in which the flux increases proportionally to the magnetizing force.

Mr. H. S. Baker has proposed the following method : * Draw a tangent to the saturation curve (at point under consideration), cutting the Y axis. The percentage of saturation is the percentage that the intercept (OT) on the Y axis is of the ordinate (ab) of the point. It will be seen that this definition is independent of the scales to which the curve is plotted, as indicated by similar points (b, b', and b'') on the saturation curves shown plotted to different scales. Thus the percentage of saturation of the point b is

$$100 \times \frac{OT}{ab} = 75.2.$$

23. LAW OF MAGNETIC CIRCUIT

Magnetomotive Force (m. m. f., symbol $\mathcal{F}$) is the total magnetizing force developed in a magnetic circuit by a coil of wire through which a current is flowing. The unit is the *Gilbert.*

$$\mathcal{F} = 0.4 \, \pi \, IN. \tag{31}$$

The magnetizing force is equal to the gilberts per centimeter length,

or $$\mathcal{H} = \frac{\mathcal{F}}{l_m}, \quad (32) \quad = \quad \frac{0.4 \, \pi IN}{l_m}, \tag{33}$$

wherein $l_m =$ mean length of magnetic circuit in centimeters.

* *Electrical World and Engineer*, Vol. XLVI, 1905, p. 1037.

What may be called the Ohm's law of the magnetic circuit is as follows: *The flux is equal to the magnetomotive force divided by the reluctance,*

$$\text{or} \qquad \phi = \frac{\mathcal{F}}{\mathcal{R}}. \qquad (34)$$

By rearrangement,

$$\mathcal{F} = \phi\mathcal{R}, \qquad (35) \qquad \mathcal{R} = \frac{\mathcal{F}}{\phi}. \qquad (36)$$

In air the reluctance is constant, and is proportional to the length of the air-gap divided by its cross-sectional area, but in any magnetic material

$$\mathcal{R} = \frac{l_m}{A\mu}, \qquad (37)$$

wherein l_m = mean length of magnetic circuit,

   A = cross-sectional area,

and μ = permeability.

Substituting values of $\mathcal{F}$ and $\mathcal{R}$ from (31) and (37) in (34),

$$\phi = \frac{0.4\,\pi IN}{\left(\dfrac{l_m}{A\mu}\right)}, \qquad (38)$$

$$\text{or} \qquad \phi = \frac{0.4\,\pi INA\mu}{l_m}, \qquad (39)$$

$$(0.4\,\pi = 1.25664).$$

$$\text{Since} \qquad \mathcal{H} = \frac{\mathcal{F}}{l_m}, \qquad (32)$$

(39) becomes $\phi = \mathcal{H}A\mu.$ $\qquad (40)$

$$\text{Now} \qquad \mathcal{B} = \frac{\phi}{A}. \qquad (41)$$

Substituting the value of $\mathscr{B}$ in (40),

$$\mathscr{B} = \mathscr{H}\mu. \qquad (42)$$

whence

$$\mathscr{H} = \frac{\mathscr{B}}{\mu}. \qquad (43)$$

Substituting the value of $\mathscr{H}$ from (33) in (42),

$$IN = \frac{0.7958\,\mathscr{B}l_m}{\mu}, \qquad (44)$$

whence

$$\mathscr{B} = \frac{0.4\,\pi IN\mu}{l_m}. \qquad (45)$$

24. PRACTICAL CALCULATION OF MAGNETIC CIRCUIT

Equation (45) shows that for any specific case the induction $\mathscr{B}$ is proportional to the ampere-turns per centimeter length of magnetic circuit. By means of the curves in Fig. 20 * the proper ampere-turns may be quickly determined, since the total number of ampere-turns required to maintain the induction $\mathscr{B}$ is equal to the product of the length of the magnetic circuit into the ampere-turns per centimeter length.

As an example, assume that 13,500 lines per square centimeter or 13.5 kilogausses are required in a wrought-iron ring, the average length of the magnetic circuit being 25 cm. Referring to Fig. 20, there are required for 13,500 lines per square centimeter 7.5 ampere-turns for each centimeter length of magnetic circuit. Hence, the total ampere-turns will be $25 \times 7.5 = 187.5$.

As a rule, the magnetic circuit consists of a uniform quality of iron. Hence, when the cross-section varies,

* From Foster's *Electrical Engineer's Pocket Book*, by permission of D. Van Nostrand Co.

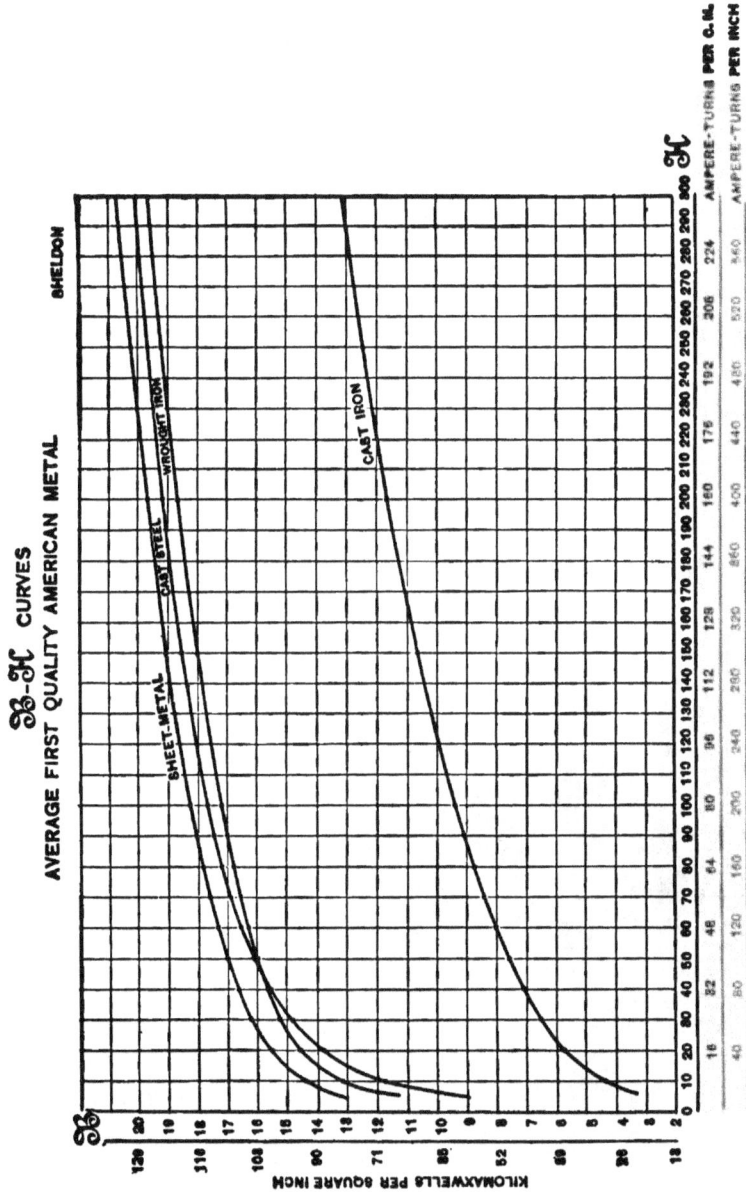

FIG. 20. — Ampere-turns per Unit Length of Magnetic Circuit.

the induction may be calculated for one part of the circuit and then, for the other parts, the induction will simply depend on the ratio of their cross-sections to the cross-section of the first part.

Where they occur, the ampere-turns for the air-gaps may be calculated by equation (44). Allowances must also be made for the curvature of the lines in air-gaps, and for joints, cracks, etc. Joints should be carefully faced, and the cross-sectional area should at least equal that of the part having the lowest permeability.

The above applies to closed magnetic circuits with small air-gaps, no leakage being considered.

25. MAGNETIC LEAKAGE

Just as electric currents in divided or branched circuits are proportional to the conductances of these circuits, so too is the number of magnetic lines flowing through iron and air in shunt with one another proportional to the *Permeances* of iron and air. As previously stated, the permeability of air is unity for all flux densities, while that of iron or steel changes with the flux density.

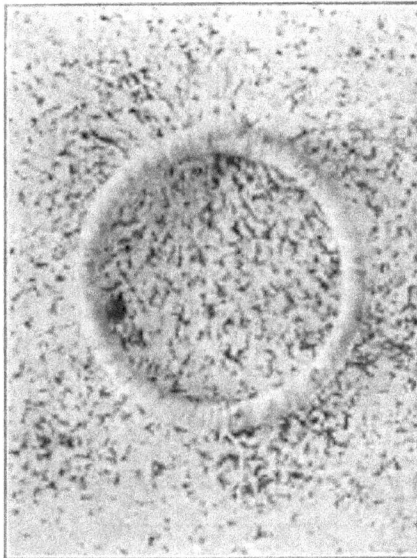

FIG. 21. — Absence of External Field.

Figures 21 to 23 show the leakage paths very nicely for an iron ring. It will be observed that in

Fig. 21, where the winding is evenly distributed over the iron ring, there are no leakage lines; whereas, in Fig. 22, in which but a small portion of the iron ring is wound with insulated wire, there is some leakage which is proportional to the relative reluctances of the iron ring and the air space. The leakage around the air-gap in Fig. 23 is very marked.

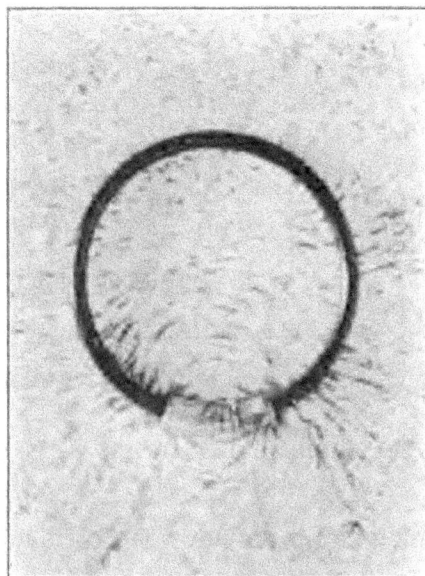

FIG. 22. — Leakage Paths.

The ratio between the total number of lines generated and the number of useful lines is called the *Leakage Coefficient*, and is denoted by the symbol V_l. Thus,

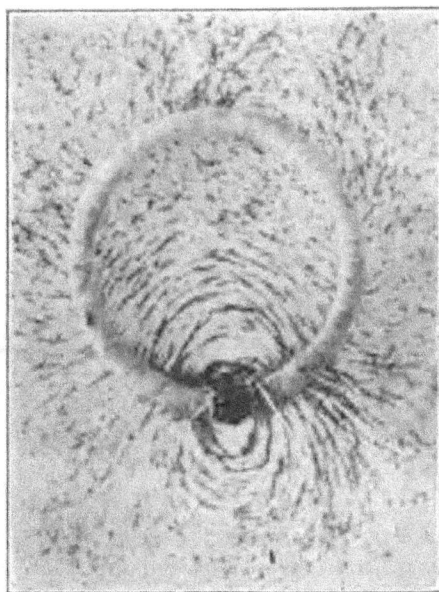

FIG. 23. — Leakage Paths around Air-gap.

$$V_l = \frac{\phi_l}{\phi_g}, \qquad (46)$$

wherein $\phi_l =$ total flux, and $\phi_g =$ useful flux through air-gap.

The reluctance between two flat surfaces is

$$\mathcal{R} = \frac{l_m}{A\mu}, \qquad (37)$$

but between two cylinders * it is

$$\mathcal{R} = \frac{0.737 \log_{10} \dfrac{a}{a_1}}{l_c}, \tag{47}$$

·wherein
$$\frac{a}{a_1} = \frac{d_c}{b - \sqrt{b^2 - d_c^2}}.$$

$$d_c = \text{diameter of the cylinders,}$$
$$b = \text{distance between centers,}$$

and
$$l_c = \text{length of cylinders.}$$

The numerical value of $\dfrac{a}{a_1}$ is constant for all dimen-

sions as long as the ratio $\dfrac{b}{d_c}$ is constant.

Figure 24 † shows the magnetic reluctance per centi-meter between two parallel cylinders surrounded by air, and having various values of the ratio $\dfrac{b}{d_c}$.

The total reluctance between two cylinders is the reluctance per centimeter divided by the length of each cylinder in centimeters. Since at the yoke the m. m. f. is approximately zero, and at the poles it is approxi-mately maximum, the average

$$\text{m. m. f.} = \frac{0.4 \, \pi IN}{2} = 0.2 \, \pi IN. \tag{48}$$

Therefore, the leakage is

$$\phi_l = \frac{0.2 \, \pi IN}{\mathcal{R}}, \tag{49}$$

* Jackson's *Electromagnetism and the Construction of Dynamos.*
† Plotted from table in Jackson's *Electromagnetism and the Con-struction of Dynamos.*

$\mathcal{R}$ being found from the curve as explained above. From this the leakage coefficient V_l is found.

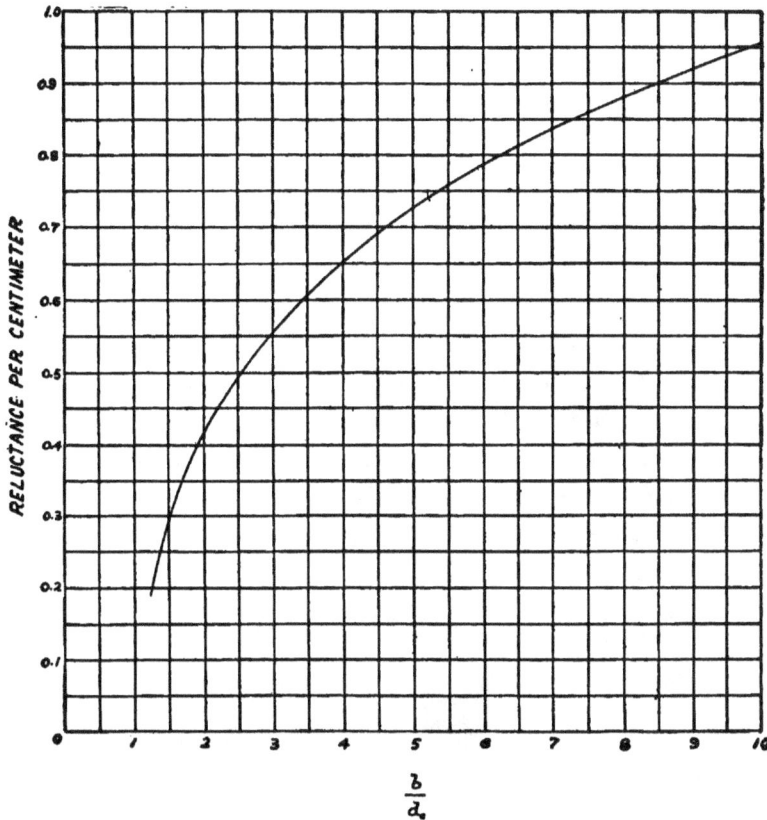

FIG. 24. — Reluctance between Cylinders.

The leakage may be included in the total reluctance by multiplying the sum of the reluctances by the leakage coefficient.

Thus, $$\mathcal{R}_{0l} = V_l(\mathcal{R}_{0_1} + \mathcal{R}_{0_2} + \mathcal{R}_{0_3} + \cdots \mathcal{R}_{0_n}). \qquad (50)$$

Only approximate results may be obtained by this method since the m. m. f. between the poles cannot be considered as the total m. m. f.

CHAPTER V

THE SOLENOID

26. DEFINITION

AN electrical conductor when wound in the form of a helix is called a *Solenoid*, and when a current of electricity is passed through the turns of wire, it possesses many of the characteristics of the bar permanent magnet, so far as its magnetic field is concerned.

As was explained in Art. 17, when a conductor carrying a current is bent into the form of a circle, the lines of force pass through the region inside of the loop. Now,

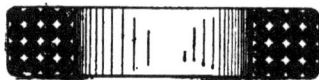

FIG. 25. — Sixteen-turn Coil.   FIG. 26. — One-turn Coil.

when several turns of wire are wound in close proximity, and a current flows through them, the lines of force due to the current in each turn unite, and the magnetic field is similar to what it would be if a solid ring of conducting material were used instead of the several turns of wire.

Referring to Figs. 25 and 26, sixteen turns of wire, carrying 100 amperes each, may be considered as equivalent to one turn carrying 1600 amperes; the ampere-turns and, consequently, the magnetic effect will be practically the same in both cases. Any coil or winding, no matter how long, or how coarse or fine the wire used, may be considered in the same light, providing the turns are not farther apart than in ordinary practice.

The internal magnetic field of a solenoid in which the length of the winding is great, as compared to its average radius, is very uniform, and this fact is taken advantage of in the type of electromagnet known as the *Coil-and-Plunger*, but more commonly called the *solenoid*.

As was explained in Art. 16, mutual attraction exists between a coil of wire carrying a current and a magnetized rod of steel. This attraction is the result of the interlinking of the lines of force due to the current in the coil, and those existing in and about the steel bar. In this case attraction or repulsion will occur, depending upon the relative directions of the magnetic fields, the bar being drawn inside the winding or repelled, as the case may be.

Magnetized steel bars cannot, however, be used, practically, to obtain repulsion, owing to the demagnetizing effect on the steel bar when the polarity of the field due to the current in the coil is reversed. Common types of solenoids have soft iron or steel *plungers*, as the bars of magnetic material are called.

The solenoid, in one of its simplest forms, consists, essentially, of the winding of helices of right and left pitch and the supporting ends, as shown in Fig. 27. Solenoids with soft iron or steel plungers are automatic

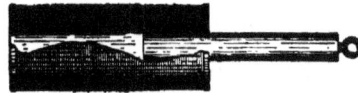

Fig. 27. — Simple Solenoid.

in their action, the plunger being magnetized by the field of the excited coil, and mutual attraction then results between the field of the coil and the induced field of the plunger, and if the coil be stationary and the plunger be free to move, the latter will be drawn within the coil until the center or neutral point of the plunger is at the center of the coil, and force will be required to change the relative positions of the coil and plunger in either direction.

27. FORCE DUE TO SINGLE TURN

In order to thoroughly understand the action of the solenoid on its plunger, consider again the loop or single turn of wire through which a current is passing. (See Art. 17, p. 26.)

Now the magnetizing force along the center line or axis of the turn is

$$\mathscr{H} = \frac{0.2\,\pi I r^2}{(r^2 + x^2)^{\frac{3}{2}}}, \tag{29}$$

wherein r is the radius as measured from the center of the wire to the axis of the loop, and x is any distance from the vertical center of the loop on its axis, in either direction. It will be seen that $\mathscr{H}$ is not only dependent upon the strength of the current, but also upon the radius of the loop and the distance x.

By reducing $0.2\,\pi I$ to unity,

$$\mathscr{H} = \frac{r^2}{(r^2 + x^2)^{\frac{3}{2}}},$$

and the characteristics of the loops and groups of loops may now be studied without considering the actual current flowing through the wire.

In Fig. 28 are shown the values for $\dfrac{r^2}{(r^2 + x^2)^{\frac{3}{2}}}$ with values of r from 1 to 14 and distances x from 0 to 10. r and x are expressed in centimeters. While the chart

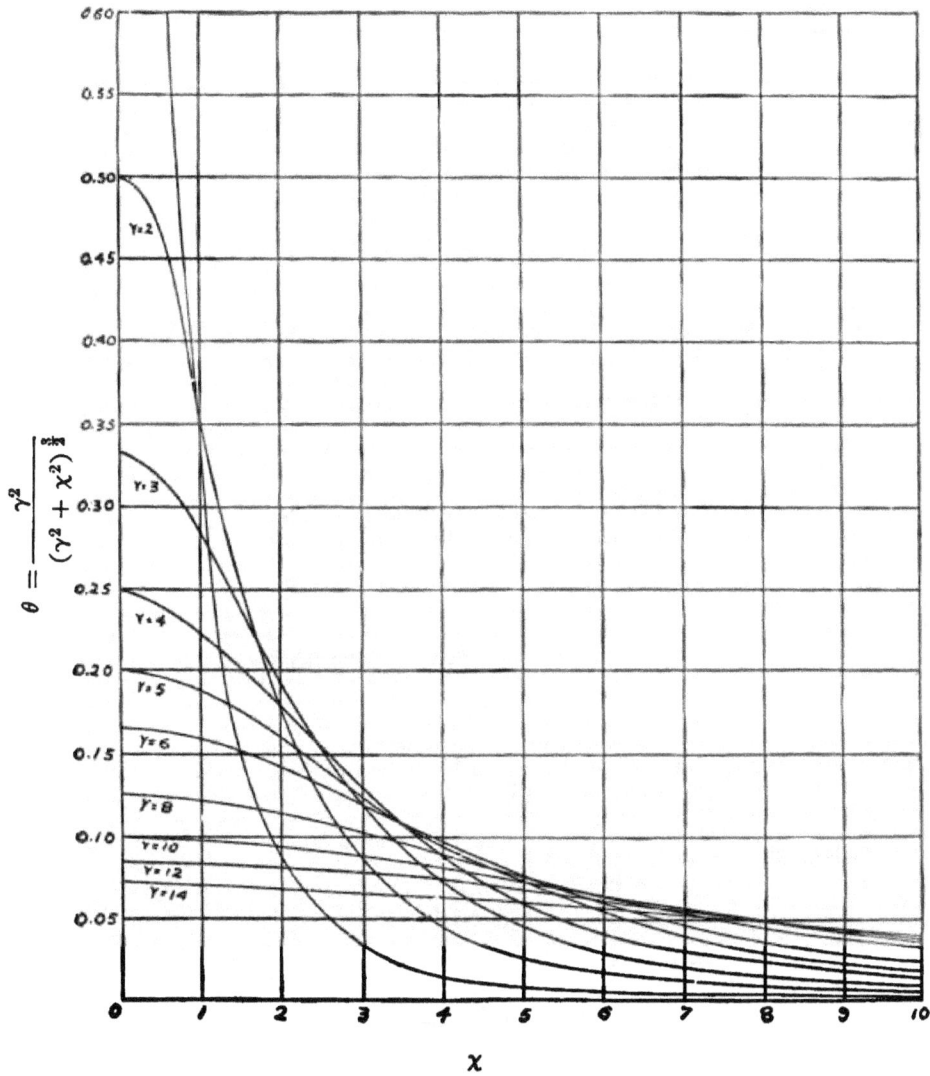

FIG. 28. — Force due to Turns of Different Radii.

only shows the curves on one side of the vertical center
line of the loop, the curves at the left-hand side would
be exactly similar, falling away from the relatively high
values at the center of the loop to lower values as the
distance x increases.

When $r = x$, $\dfrac{r^2}{(r^2 + x^2)^{\frac{3}{2}}} = \dfrac{0.3536}{x}$. The factor 0.3536
is merely $\dfrac{1}{2^{\frac{3}{2}}}$. It will also be observed that when $r = x$,
the curve for twice the value of r, that is, $2\,r$, intersects
with r and x at nearly the same point. Thus, the $r = 6$
curve intersects with the $r = 3$ curve at $x = 3$. This is
due to the fact that when $r = x$,

$$\frac{r^2}{(r^2 + x^2)^{\frac{3}{2}}} = \frac{r^2}{(2\,r^2)^{\frac{3}{2}}} = 0.3536\ \sqrt{r},$$

and when $r = 2\,x$,

$$\frac{r^2}{(r^2 + x^2)^{\frac{3}{2}}} = \frac{(2\,x)^2}{((2\,x)^2 + x^2)^{\frac{3}{2}}} = 0.358\sqrt{r}.$$

Assigning θ to $\dfrac{r^2}{(r^2 + x^2)^{\frac{3}{2}}}$, these relations may be expressed $\theta = \dfrac{0.3536}{(r = x)}$ and $\theta = \dfrac{0.358}{\left(\dfrac{r}{2} = x\right)}$. Hence, if
$r = 6$, at $\dfrac{6}{2} = 3$ cm. from the vertical center of the loop
$(= x)$ the value of θ will be $\dfrac{0.358}{3} = 0.1193$; or if
$r = 3 = x$, $\theta = \dfrac{0.6536}{3} = 0.1179$.

It is seen that where the radius r is small, the force
along the horizontal axis is great for small values of x,
whereas with the larger radii, the force is more uniform,

and is greater for large values of x than for the turns of smaller radii.

The equation $\theta = \dfrac{0.3536}{(r=x)}$ shows that a curve drawn through the various points of intersection of r and x, where $r = x$, would be a rectangular hyperbola, which indicates that the total work done in moving a unit magnetic pole from an infinite distance to the vertical center line of the turn of wire along the horizontal axis would be the same in all cases.

28. FORCE DUE TO SEVERAL TURNS ONE CENTIMETER APART

The practical solenoid consists of many layers of wire, each layer consisting of a great many turns. Hence, it is necessary to know what relation exists between a single turn and a great many turns.

Consider two turns of wire of 2 cm. radius, arranged side by side and 1 cm. apart, center to center, with the same amount of current flowing through each.

Referring to the chart, Fig. 28, it is seen that at the center of each turn of wire $\theta = 0.5$, and at 1 cm. on each side $\theta = 0.358$. It is obvious, then, that the total force at the center of each turn will be the sum of the forces due to each turn, and this will be proportional to $0.5 + 0.358 = 0.858$.

Likewise, if another turn be added under similar conditions, the force may be determined as follows:

TURN 1

θ at center	0.500
θ due to turn 2, 1 cm. distant . . .	0.358
θ due to turn 3, 2 cm. distant . . .	0.117
$\theta_{L1} =$ total force for turn 1	1.035

TURN 2

θ at center	0.500
θ due to turn 1, 1 cm. distant . . .	0.358
θ due to turn 3, 1 cm. distant . . .	0.358
θ_{L_2} = total force for turn 2	1.216

TURN 3

θ at center	0.500
θ due to turn 2, 1 cm. distant . . .	0.358
θ due to turn 1, 2 cm. distant . . .	0.177
θ_{L_3} = total force for turn 3	1.035

In the above, θ_L indicates the force due to several turns arranged side by side, all being of the same radius and θ_{L_1}, etc., indicates the number of the turn.

Adding one more turn, the following is obtained:

TURN 1

θ at center	0.500
θ due to turn 2, 1 cm. distant . . .	0.358
θ due to turn 3, 2 cm. distant . . .	0.177
θ due to turn 4, 3 cm. distant . . .	0.085
θ_{L_1} = total force for turn 1	1.120

TURN 2

θ at center	0.500
θ due to turn 1, 1 cm. distant . . .	0.358
θ due to turn 3, 1 cm. distant . . .	0.358
θ due to turn 4, 2 cm. distant . . .	0.177
θ_{L_2} = total force for turn 2	1.393

Turn 3

θ at center	0.500
θ due to turn 2, 1 cm. distant . . .	0.358
θ due to turn 4, 1 cm. distant . . .	0.358
θ due to turn 1, 2 cm. distant . . .	0.177
θ_{L3} = total force for turn 3	1.393

Turn 4

θ at center	0.500
θ due to turn 3, 1 cm. distant . . .	0.358
θ due to turn 2, 2 cm. apart	0.177
θ due to turn 1, 3 cm. distant . . .	0.085
θ_{L4} = total force for turn 4	1.120

This procedure may be continued indefinitely.

A little reflection will show that the sums of the values of θ may be expressed in the following arrangement, wherein θ_1 represents the value of θ for the first (central) turn, θ_2 the value of θ for the second (adjoining) turn, etc.

No. of
Turns

1. θ_1.
3. $\theta_2 + \theta_1 + \theta_2$.
5. $\theta_3 + \theta_2 + \theta_1 + \theta_2 + \theta_3$.
7. $\theta_4 + \theta_3 + \theta_2 + \theta_1 + \theta_2 + \theta_3 + \theta_4$.
9. $\theta_5 + \theta_4 + \theta_3 + \theta_2 + \theta_1 + \theta_2 + \theta_3 + \theta_4 + \theta_5$.

If N_c = the number of turns, or groups of turns, of wire, m = the number of the central turn or group, and θ_{Lm} = the sum of the values of θ = total θ at the center

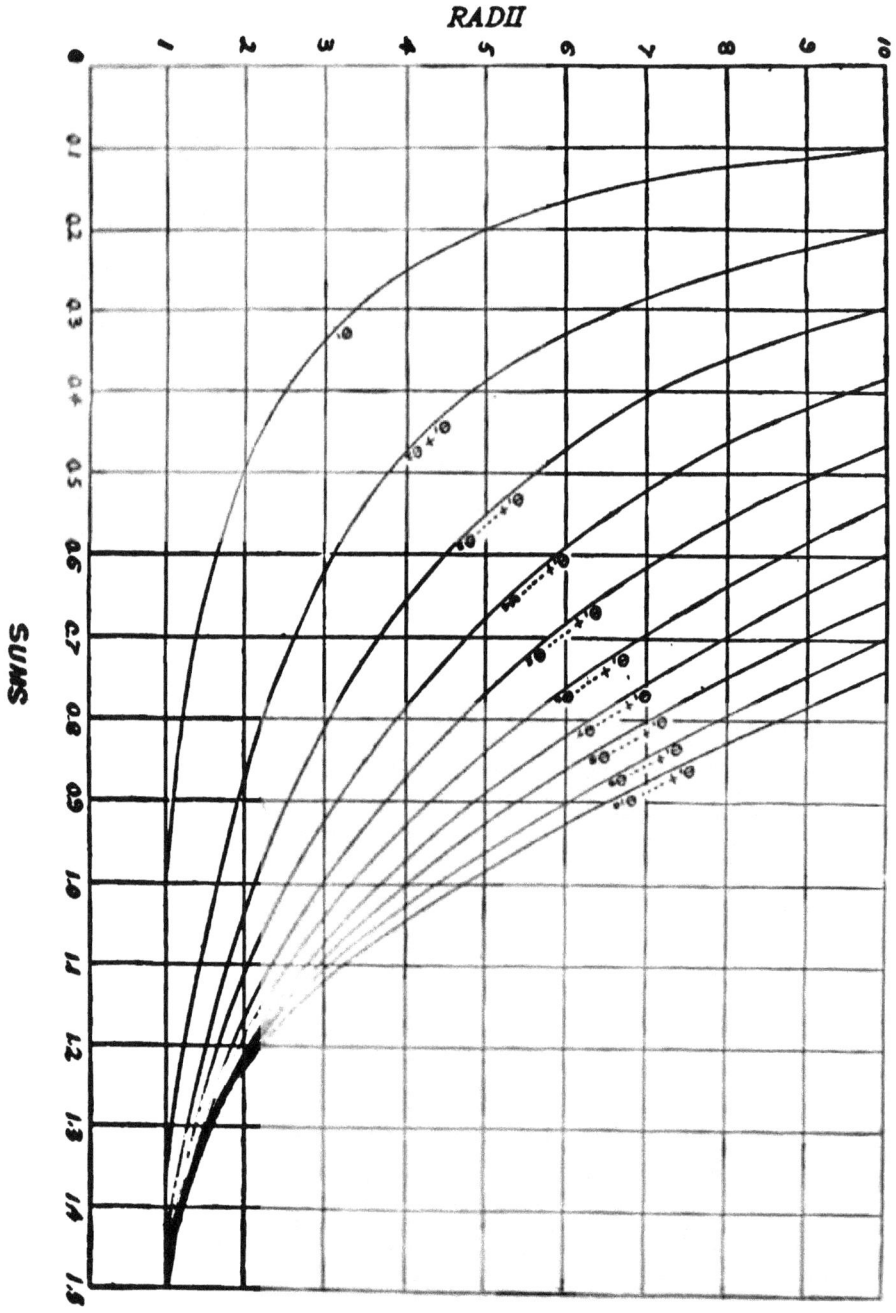

RADII

SUMS

FIG. 29.—Sums of Forces for Various Radii of Turns.

of the coil when the turns or groups are 1 cm. apart, center to center,

then $\qquad \theta_{Lm} = 2\,(\theta_1 + \cdots \theta_m) - \theta_1.$ \hfill (51)

The sums of θ_1 to θ_{10}, for various radii, are shown in Fig. 29.

29. FORCE DUE TO SEVERAL TURNS PLACED OVER ONE ANOTHER

A method of calculating the force at the center of a coil of any length and radius (this will be the average radius in the case of any wire or group of wires larger than a point) with unit thickness or depth, T (in the case of groups of wires), has been given. It now remains to calculate the total force for solenoids of any thickness and any radius. It is obvious that if the successive addition of the forces due to adjoining turns gives the total force, so should the addition of the forces due to several turns or groups placed one over another give the total force at the center due to all the turns.

Consider the arrangement in Fig. 30. Each square may be assumed to represent either a solid conductor or a group of smaller conductors insulated from one another, the total ampere-turns being the same in each square.

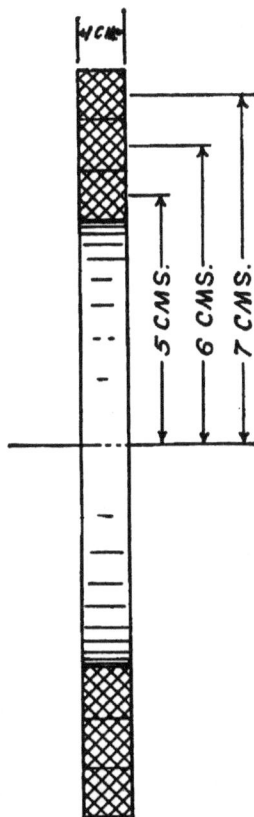

Fig. 30.—Groups of Turns placed over One Another.

At the center of a loop of wire

$$\mathscr{H} = \frac{0.2\,\pi I}{r}, \tag{25}$$

or

$$\mathscr{H} = \frac{0.2\,\pi IN}{r}, \tag{52}$$

where there is more than one turn, and since, in this case, $x = 0$, $\theta = \dfrac{r^2}{(r^2+x^2)^{\frac{3}{2}}} = \dfrac{r^2}{r^3} = \dfrac{1}{r}$.

The sum of the forces at the center of the coil in Fig. 30 will, therefore, be

$$\theta_v = \frac{1}{5} + \frac{1}{6} + \frac{1}{7} = \frac{42+35+30}{210} = \frac{107}{210} = 0.51.$$

The magnetizing force at the center may also be calculated direct from the *mean magnetic radius* of the coil in Fig. 30, which is known as a *disk solenoid* or *disk winding*.

The mean magnetic radius is equal to the reciprocal of the sum of the reciprocals of the average radii of the squares or groups of turns constituting the disk, multiplied by the number of squares.

Now the average radius r_a is 6, but the mean magnetic radius is

$$r_m = \frac{N_g(r_c r_d r_e)}{(r_c r_d) + (r_d r_e) + (r_e r_c)}, \tag{53}$$

wherein N_g is the number of groups, and r_c, r_d, and r_e, etc., are the respective average radii of the groups of turns.

Hence, in the case considered,

$$r_m = \frac{3(5 \times 6 \times 7)}{(5 \times 6) + (6 \times 7) + (7 \times 5)} = \frac{630}{107} = 5.9,$$

and since $\theta_v = \dfrac{1}{r}$ for one group, for the three groups

$$\theta_v = \frac{N_g}{r_m}\left(54\right) = \frac{3}{5.9} = 0.51,$$

as in the previous case.

30. Force due to Several Disks placed Side by Side

Having found the relations for the disk winding, consider the effect of placing several of the disks side by side, as in the case of the loops of the same radii. For this purpose the disk winding in Fig. 30 will be selected, since its constants have already been calculated.

Arranging three such disks side by side, which will make the distance apart, center to center, one centimeter, they will appear as in Fig. 31.

Substituting T for N_g, which expresses the vertical thickness of the winding, and assigning L to the length of the winding, the values will be $T = 3$, $L = 3$.

Calculating on the same principle as used in the determination of the force at the center of a group of single loops placed side by side, it is

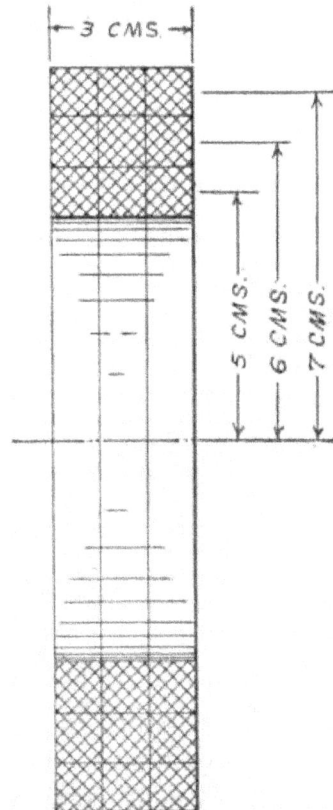

Fig. 31. — Groups of Turns arranged to form a Large Square Group.

obvious that the force at any point x on the axis of a single disk will be

$$\theta_{v_m} = \frac{T\, r_m{}^2}{(r_m{}^2 + x^2)^{\frac{3}{2}}}.$$ (55)

Hence, when $x = 1$,

$$\theta_{v_1} = \frac{3 \times 34.67}{(1 + 34.67)^{\frac{3}{2}}} = \frac{104}{212.8} = 0.489,$$

while at the center of each disk, $\theta_v = 0.51$.

Following the same general procedure as given on p. 50, there results for the total force θ_s at the center of the entire square winding,

$$\theta_s = \theta_v + \theta_{v_1} + \theta_{v_2} = 0.51 + 0.489 + 0.489 = 1.488.$$

In practice it is customary to express the magnetizing force in terms of ampere-turns per centimeter length, since $\mathcal{H}$ is proportional to $\dfrac{\mathcal{F}}{l_m}$, wherein $\mathcal{F}$ is the magnetomotive force (m. m. f.), and l_m is the mean length of the magnetic circuit in centimeters, and regardless of the thickness of the winding, T. Hence, in this case the force at the center is

$$\theta_t = \frac{\theta_s}{T}\ \text{(56)},\quad = \frac{1.488}{3} = 0.496,$$

wherein θ_t is the factor to be multiplied by $0.2\,\pi IN$ to give the value of $\mathcal{H}$ at the center of the winding.

The same result may, of course, be obtained by calculating the force due to three coils of average radii 5 cm., 6 cm., and 7 cm., respectively, each being 3 cm. in length. This gives the following result:

θ_L due to inner coil 0.578

θ_L due to middle coil 0.487

θ_L due to outer coil 0.420

$\theta_s =$ total at center $= 1.485$

Hence, $\theta_t = \dfrac{\theta_s}{T} = \dfrac{1.485}{3} = 0.495.$

While so exact values cannot be read from a chart as may be obtained by direct calculation, the former are near enough in practice. The value of the charts will be appreciated when it is considered that the formula for a simple coil where $L = 2$ and $T = 3$ would be

$$\theta_s = \frac{1}{r_{a_1}} + 2\left[\frac{r_{a_1}^2}{(1+r_{a_1}^2)^{\frac{3}{2}}}\right] + \frac{1}{r_{a_2}} + 2\left[\frac{r_{a_2}^2}{(1+r_{a_2}^2)^{\frac{3}{2}}}\right]$$

$$+ \frac{1}{r_{a_3}} + 2\left[\frac{r_{a_3}^2}{(1+r_{a_3}^2)^{\frac{3}{2}}}\right], \quad (57)$$

or

$$\theta_s = \frac{1}{r_{a_1}} + \frac{1}{r_{a_2}} + \frac{1}{r_{a_3}} + 2\left[\left[\frac{r_{a_1}^2}{(1+r_{a_1}^2)^{\frac{3}{2}}}\right] + \left[\frac{r_{a_2}^2}{(1+r_{a_2}^2)^{\frac{3}{2}}}\right]\right.$$

$$\left.+ \left[\frac{r_{a_3}^2}{(1+r_{a_3}^2)^{\frac{3}{2}}}\right]\right], \quad (58)$$

wherein r_{a_1}, r_{a_2}, and r_{a_3} are the radii of the inner, middle, and outer squares or groups respectively.

31. Force at Center of any Winding of Square Cross-section

Since the rule $\theta = \dfrac{1}{r}$ holds for a single loop or group of turns one square centimeter in cross-section, it is obvious that a slight modification of this rule should apply to the total force at the center of any winding of square cross-section.

This relation may be expressed $\theta_t = \dfrac{L}{r_a}$ (59) (nearly), wherein L is the length, and r_a the average radius of the winding.

Applying this formula to the coil in Fig. 31, there results $\theta_t = \frac{3}{6} = 0.5$.

This type of solenoid is sometimes called a *rim* solenoid in contradistinction to the disk solenoid.

32. Tests of Rim and Disk Solenoids

The following tests * which were made by the author will be of interest in proving the foregoing formulæ.

Two sets, each consisting of four flat spools or bobbins of varying outside diameters, were prepared with a hole in the center of each large enough to receive a plunger 2.87 cm. in diameter. The cross-sectional area of the plunger was, therefore, 6.45 sq. cm.

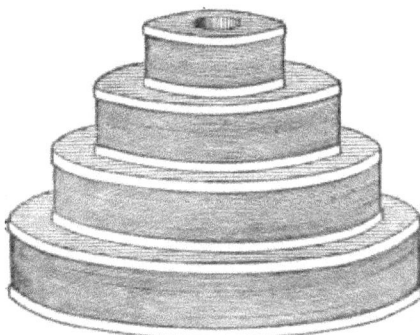

Fig. 32. — The Test Solenoids.

Figure 32 shows the

* *Electrical World and Engineer*, Vol. XVI, 1905, pp. 615–617.

general appearance of each set, while the actual dimensions of the rim and disk solenoids are shown in Figs. 33 and 34.

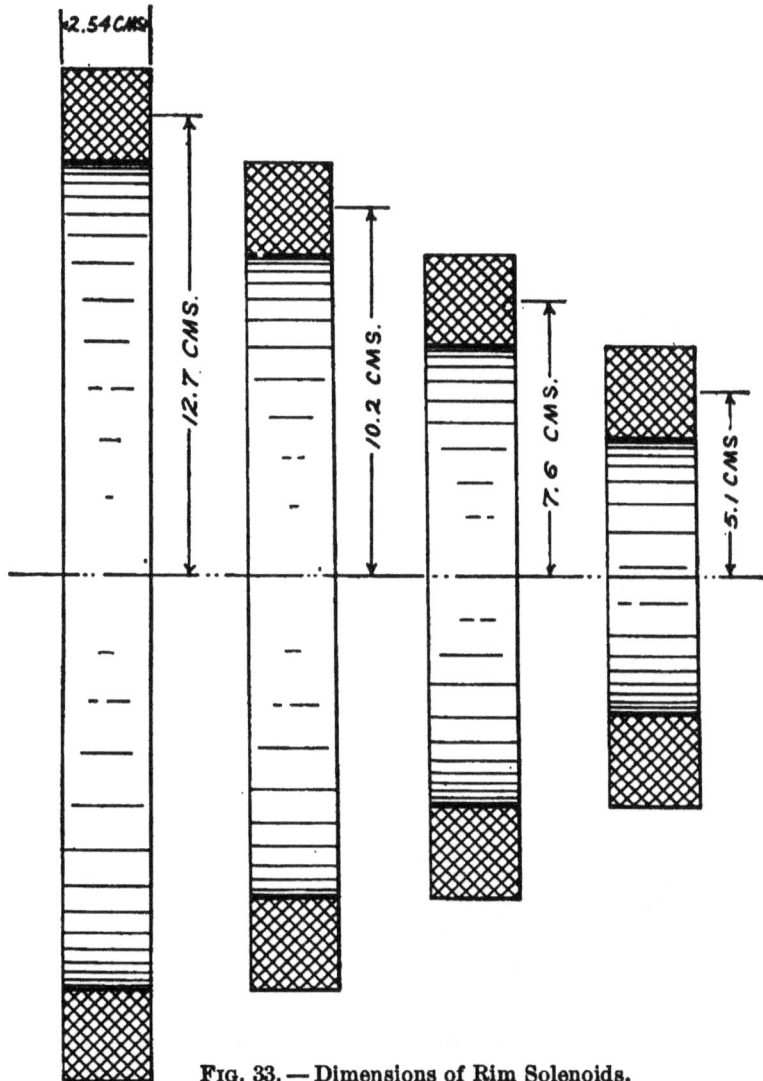

FIG. 33. — Dimensions of Rim Solenoids.

The length of the plunger was one meter. Of course, a shorter plunger could have been used, but it was desired to have a plunger of sufficient length to make the conditions of the test as ideal as possible. The cross-

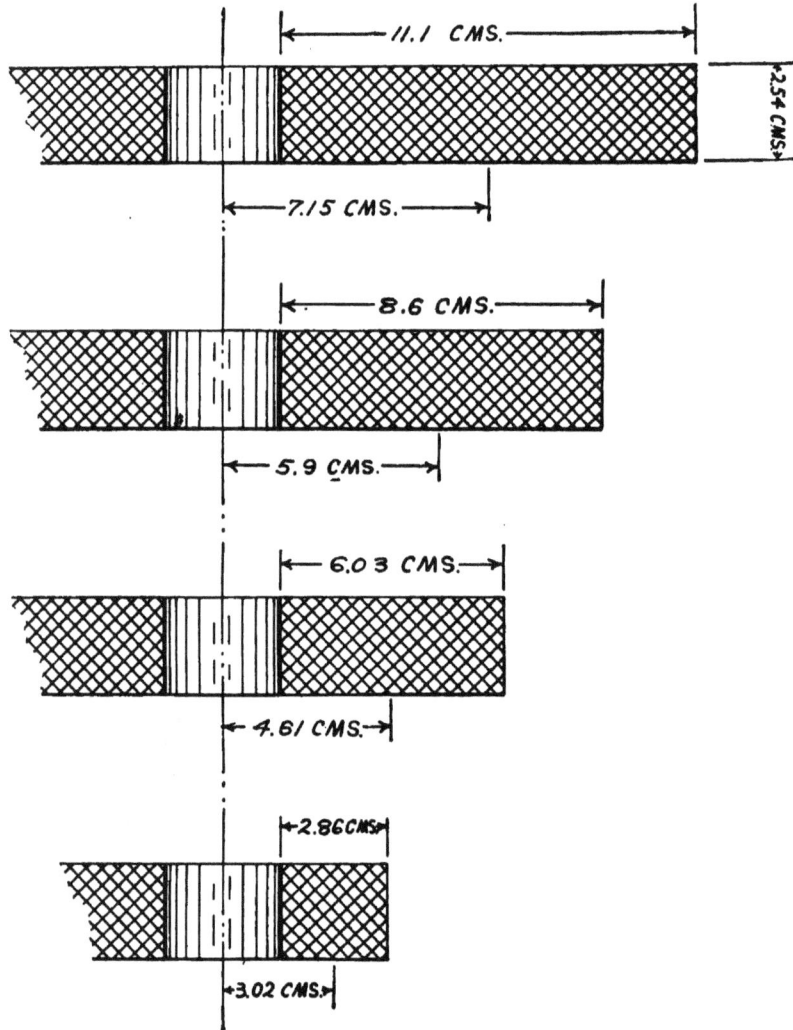

FIG. 34. — Dimensions of Disk Solenoids.

sectional area of the plunger was purposely made the same as the cross-sectional area of the rim solenoids.

The tests were made by the method illustrated in Fig. 35.   Here A is a magnetizing coil to saturate the core, and B represents the winding to be tested.

First the coil A was excited so as to thoroughly saturate the iron core, and then a disk winding was

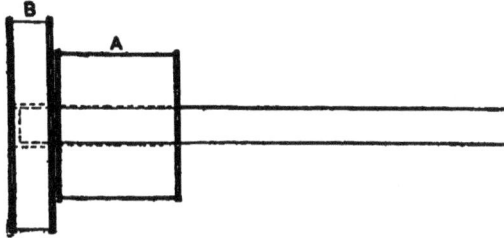

FIG. 35. — Method of testing Rim and Disk Solenoids.

placed over the end of the core in the position of maxi-
mum pull, and excited from a separate source.   Each coil had a rheostat and ammeter in series, the two coils being so connected that attraction would result.   Coil A was rigidly fastened to the iron core, and the test coil B attached to the scales.

After the core was saturated, a change in the strength of the current in coil A produced an almost inappreciable change in the pull, which was accredited to the change in $\mathcal{H}$; but when the current strength in coil B was changed, the pull varied directly with the current in coil B.

Referring to Fig. 33, it is seen that for the rim sole-
noids, $T = L = 2.54$ cm. in each case.   The values of r_a are 5.1, 7.6, 10.2, and 12.7 cm., respectively.

By using the formula $\theta_s = \dfrac{L}{r_a}(60)$, the following results are obtained:

r_a					θ_s
5.1	.	.	.	.	.	0.498
7.6	.	.	.	.	.	0.334
10.2	.	.	.	.	.	0.248
12.7	.	.	.	.	.	0.200

Now the pull due to a solenoid on its plunger is directly proportional to the strength of the current after the plunger is saturated. Figure 36 shows the relative pulls with varying degrees of excitation in the rim solenoids, and Fig. 37 shows the results of the test of the disk solenoids.

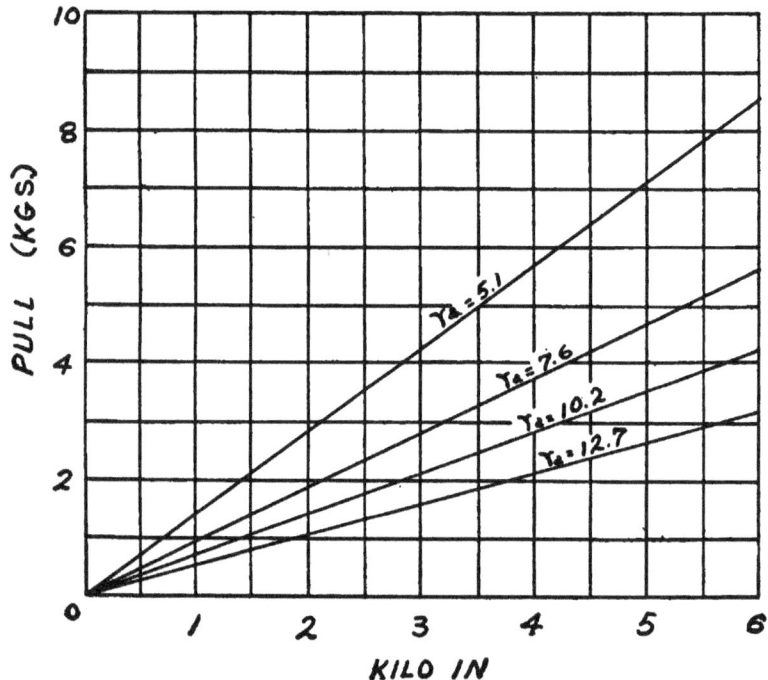

Fig. 36.— Characteristics of Rim Solenoids.

Comparing the pull P (in kilograms), for 1500 ampere-turns in each case, with the calculated θ_s values:

r_a	θ_s	P	$\dfrac{P}{\theta_s}$
5.1 . .	0.498 . .	2.16 . .	3.48
7.6 . .	0.334 . .	1.45 . .	3.42
10.2 . .	0.248 . .	1.08 . .	3.30
12.7 . .	0.200 . .	0.85 . .	3.22

It will be observed that the ratios between θ_s and P vary slightly, the relative pulls being greater for small values of r_a.   This is due to the effect of $\mathcal{H}$ in the coil used to saturate the iron core or plunger.

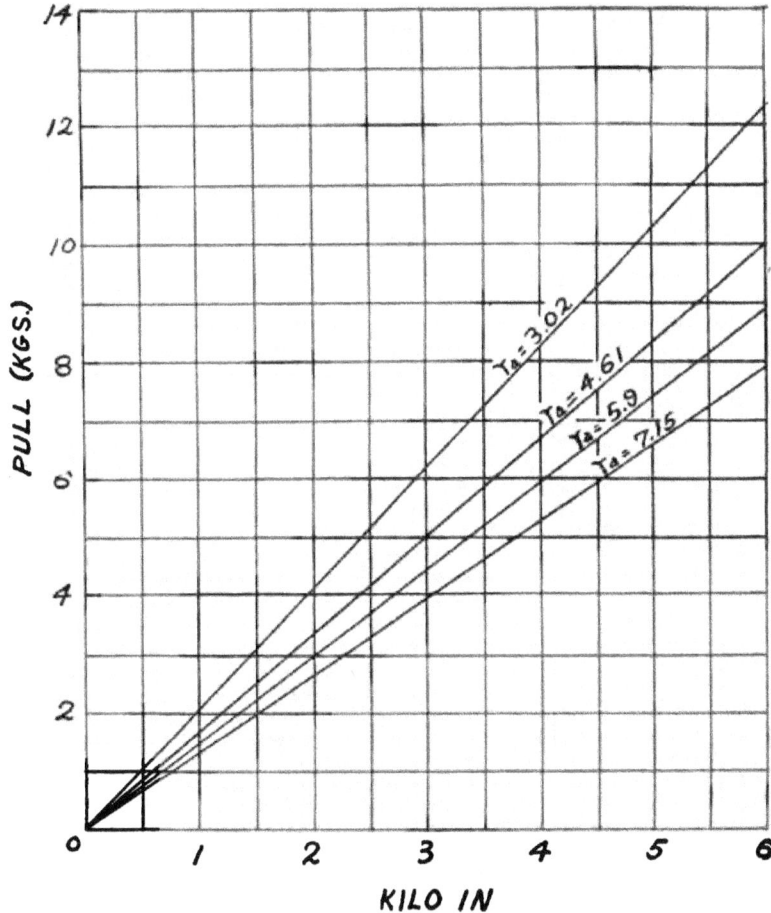

FIG. 37. — Characteristics of Disk Solenoids.

The curve from a to b in Fig. 38 is plotted from Fig. 36, the points being taken from the pulls corresponding to the different mean magnetic radii on the ordinate representing 6000 ampere-turns.

The rim windings were so dimensioned that when telescoped (theoretically), they would form a disk winding, as in Fig. 39. Now the sum of the pulls, in kilograms, due to the four rim windings, with 1500

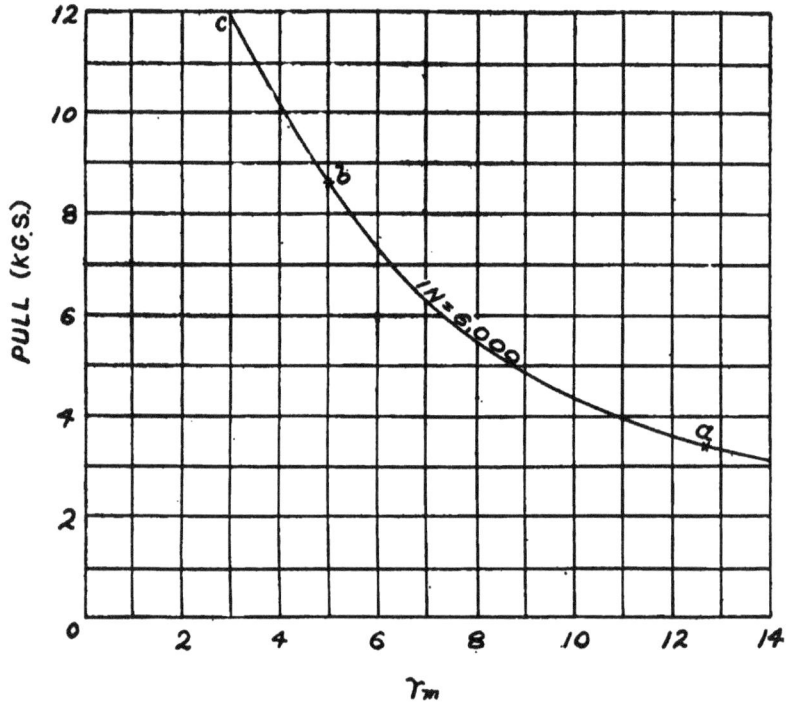

FIG. 38.—Ratio of r_m to Pull for Rim and Disk Solenoids.

ampere-turns in each winding, making a total of 6000 ampere-turns, is $2.16 + 1.45 + 1.08 + 0.85 = 5.54$ kg.

The mean magnetic radius, r_m, of the disk winding consisting of the telescoped rim windings is, according to (53),

$$\frac{4(5.1 \times 7.6 \times 10.2 \times 12.7)}{(5.1 \times 7.6 \times 10.2) + (7.6 \times 10.2 \times 12.7) + (10.2 \times 12.7 \times 5.1) + (12.7 \times 5.1 \times 7.6)} = 8.$$

Hence, since there are four sections,

$$\theta_s = \frac{4L}{r_m} = \frac{4 \times 2.54}{8} = 1.27.$$

Referring again to Fig. 38, it is seen that when $r_m = 8$, $P = 5.54$, which coincides with the results obtained with the rim windings.

By calculating the values of r_m for the disk windings after this manner, the rest of the curve a–c is obtained, which coincides with, and forms a continuation of, the curve representing the test of the rim windings.

The curve in Fig. 40 represents the product of the pulls multiplied by the mean magnetic radii, the products below $r_m = 3$ being assumed from the natural slope of the curve.

33. MAGNETIC FIELD OF PRACTICAL SOLENOIDS

In order to actually see, so to speak, just what relation exists between the lines of force due to the current in the coil and the lines of induction in the plunger,* the author made the photographs shown in Figs. 41 to 44, by placing the solenoid and plunger in hard sand, so

* *Electrical World and Engineer,* Vol. XV, 1905, p. 797.

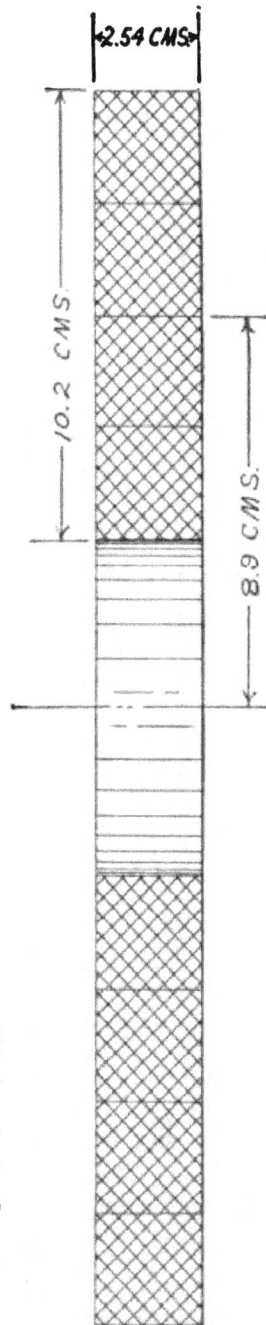

FIG. 39. — Rim Solenoids telescoped to form Disk Solenoid.

that the plane of the surface of the sand cut the center of the solenoid and plunger. The solenoid was then

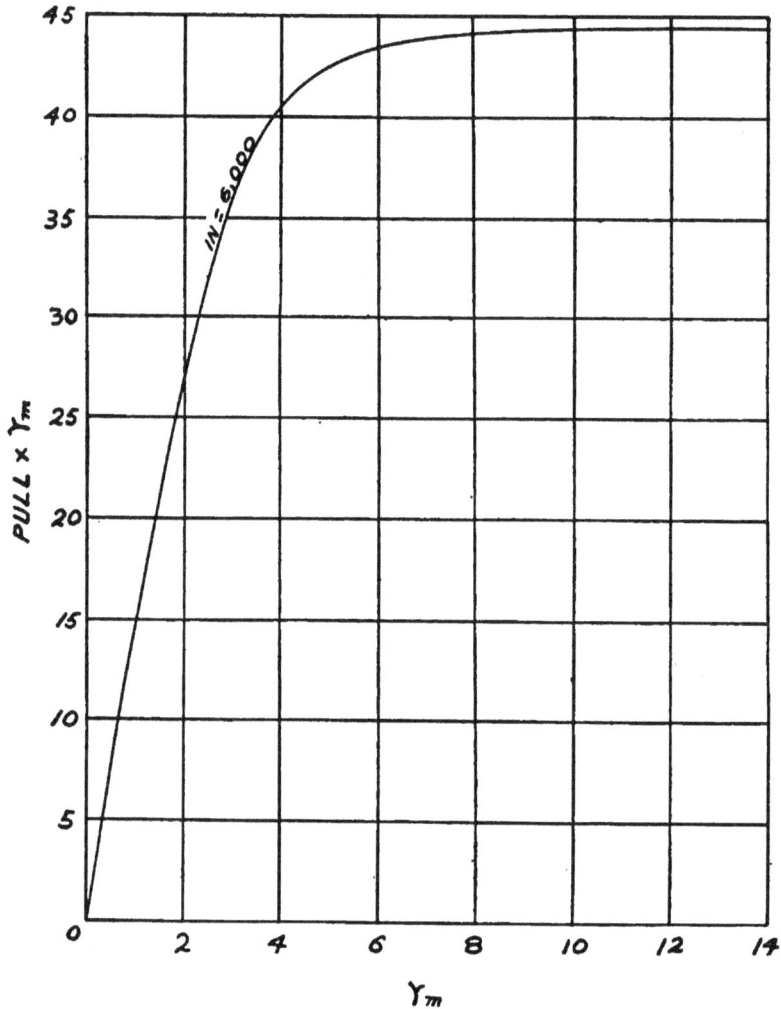

FIG. 40. — Product of Pull and Mean Magnetic Radius.

excited and iron filings sprinkled on to show the flux paths. The solenoid was equally excited in each case.

In Fig. 41 the end of the core is flush with the end of the winding. It will be observed that the field about the solenoid is very weak, and furthermore that the iron plunger appears to have but one polar region, *i.e.* the lines of force appear to leave it uniformly above the mouth of the solenoid. Hence, the polar region at the mouth of the solenoid must be very small indeed.

Figure 42 shows the plunger inserted one third of its length into the winding. The magnetic field is greatly increased, due to

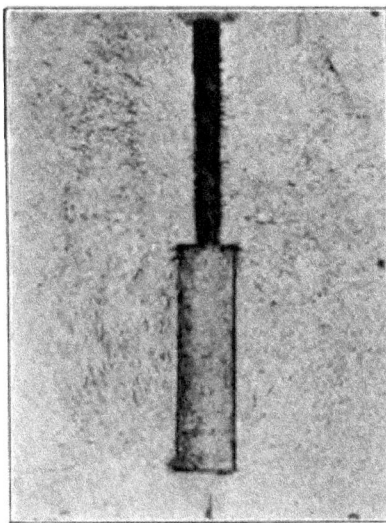

FIG. 41.—Plunger removed from Solenoid.

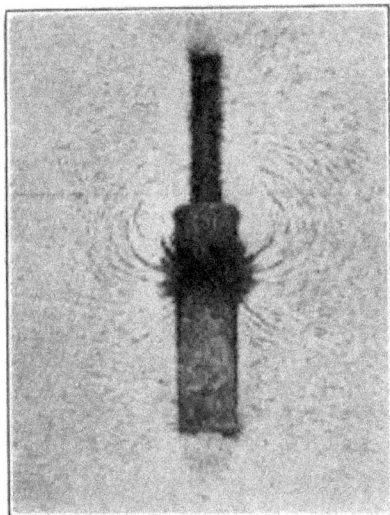

FIG. 42.—Plunger inserted one third into Solenoid.

the induction in the iron. It is very evident that there is a well-defined pole within the solenoid, but the lines of force leave the projecting portion of the plunger uniformly, indicating that the polar region is widely distributed.

The same general conditions exist in Fig. 43 as in Fig. 42, with the exception that, as the plunger is two thirds of its length within the solenoid, the field about the solenoid is stronger, and the polar surface of the plunger protruding from the solenoid is smaller.

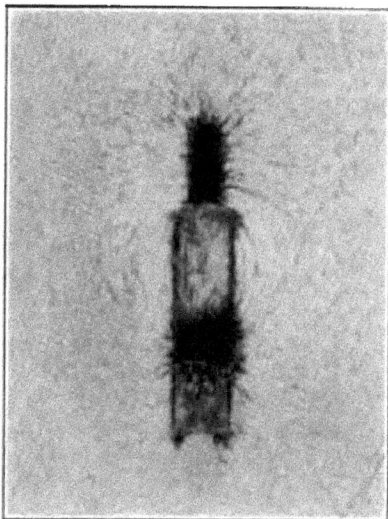

FIG. 43. — Plunger inserted two thirds into Solenoid.

The general characteristics of the bar permanent magnet are met, however, when the plunger is entirely within the solenoid, as in Fig. 44. Here the plunger will remain at rest, and if forcibly moved in either direction, it will return to its position of equilibrium.

Referring again to Figs. 42 and 43, it is evident that the position of maximum pull will be at a point between the positions shown, and such is the case for solenoids of the general dimensions of the one in question, while for very short solenoids, and for low-flux densities in the plungers, the position of maximum pull may not be reached until the end of the plunger has protruded a short distance from the solenoid.

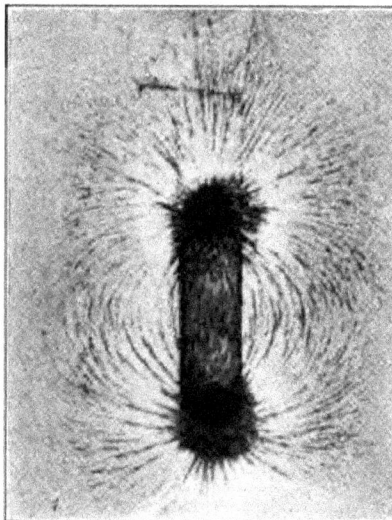

FIG. 44. — Plunger entirely within the Solenoid.

34. RATIO OF LENGTH TO AVERAGE RADIUS

By building up (theoretically) coils of various lengths, average radii, and thicknesses from the formula given, the relations between the above dimensions may be investigated.

Figure 45 shows the values of θ_t for various values of r_a and L when $T = 1$.

It will be remembered that θ_t is the force at the center of the entire coil for one centimeter of length, regardless of the thickness of the winding, since

$$\theta_t = \frac{\theta_s}{T}. \tag{56}$$

Hence, if the value of θ_t may be determined for coils of any dimensions, the magnetizing force at the center of the coil will be $\mathcal{H} = 0.2\,\pi IN\theta_t$ (61). (See p. 42.)

In Fig. 46 is shown the effect of changing the thickness of the winding. It will be seen that there is a slight variation in the values of θ_t for given values of r_a and L, for small values of r_a, but that the difference for relatively large values of r_a is inappreciable.

It will also be noticed from the slope of the curve for $r_a = 1$, that the value of θ_t gradually approaches 2, which value it can never exceed in any coil when $0.2\,\pi IN = 1$, which is the basis upon which these calculations have been made.

Referring again to Fig. 46, it will be observed that for coils of one centimeter length the relation is approximately

$$\theta_t = \frac{L}{r_a}, \tag{59}$$

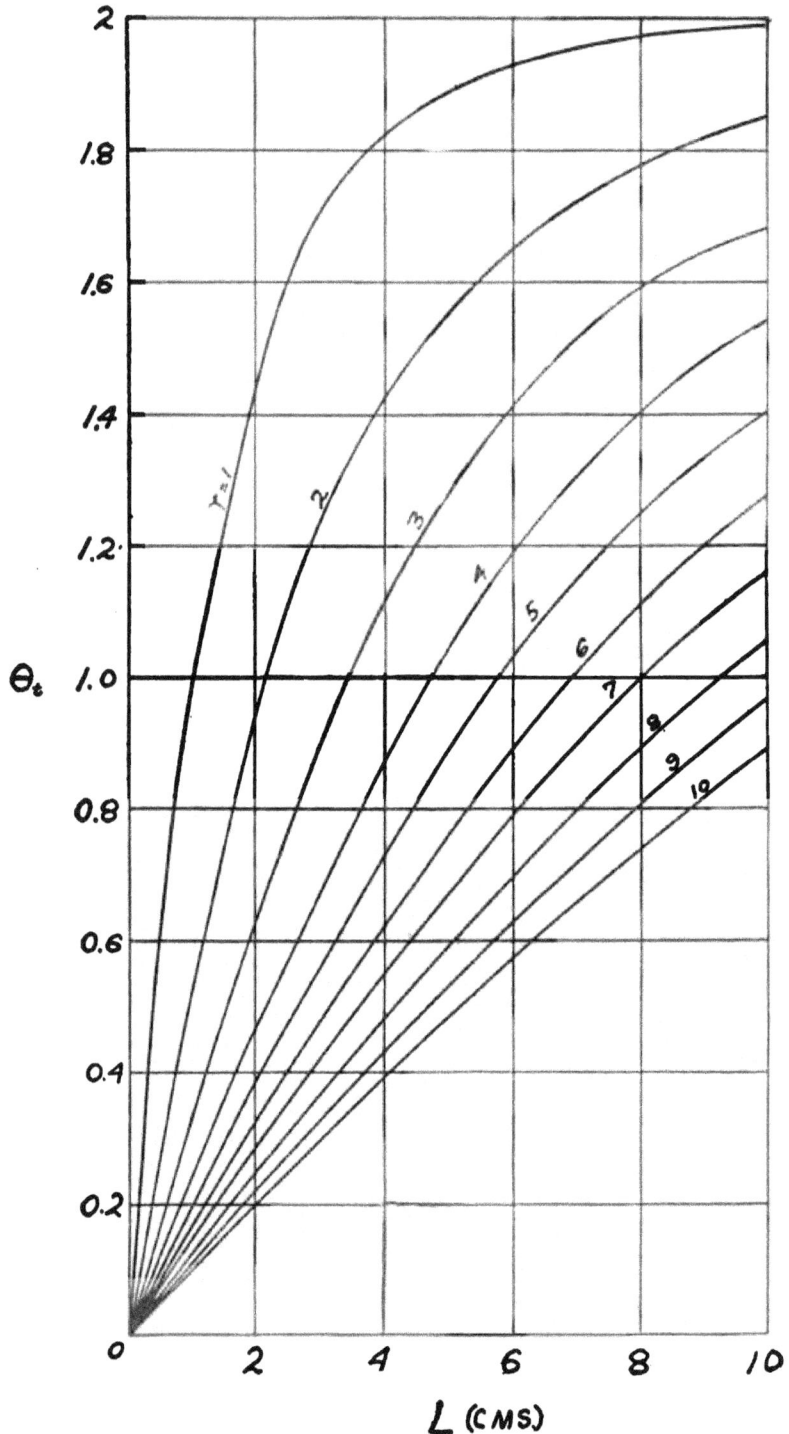

FIG. 45.—Force due to Solenoids with Unit Thickness or Depth of Winding.

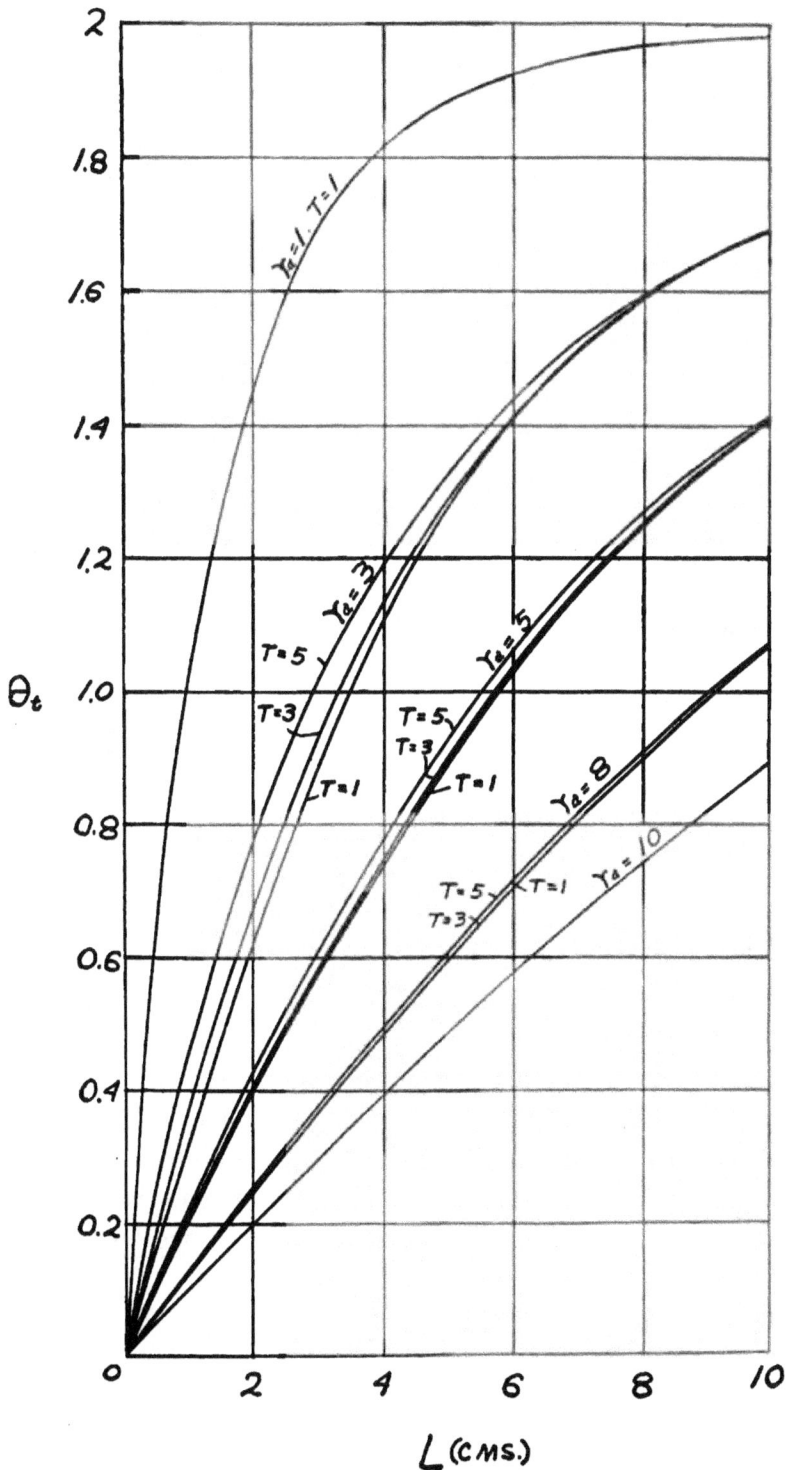

FIG. 46. — Effect of changing Thickness of Winding.

when $\theta_t = 1$ or less, while for the same values of θ_t,

$$\theta_t = \frac{0.9\,L}{r_a} \qquad (62)$$

(approximately), when L exceeds 5.

The value of θ_t above $\theta_t = 1$ may be expressed with a fair degree of accuracy by the empirical formula

$$\theta_t = 2 - \frac{r_a}{1.07\,L}. \qquad (63)$$

It is obvious that the value of θ_t, as given in (63), will constantly decrease as the length of the winding increases. Hence, for a very long solenoid of small radius, $\theta_t = 2$.

In practice the total ampere-turns are calculated direct for the entire coil. (See p. 34.) Hence, to determine the ampere-turns per centimeter length, the total ampere-turns must be divided by the length of the winding. Therefore,

$$\mathcal{H} = \frac{0.2\,\pi I N \theta_t}{L}, \qquad (64)$$

and when $\theta_t = 2$,

$$\mathcal{H} = \frac{0.4\,\pi I N}{L}, \qquad (65)$$

which is the formula for a very long solenoid of small radius, or for a coil whose core forms a closed ring.

Formula (65) is simply the m. m. f., which is always $\mathcal{G} = 0.4\,\pi I N$, divided by the length of the winding which, in the two cases mentioned above, represents the mean length l_m of the magnetic circuit.

CHAPTER VI

PRACTICAL SOLENOIDS

35. Tests of Practical Solenoids

In order to obtain practical data on the action of
solenoids, the author made numerous tests* of solenoids
of various dimensions. Five solenoids were constructed, each having an average radius of 2.76 cm.,
while the lengths were 7.63, 15.25, 22.8, 30.5, and 45.8
cm., respectively.

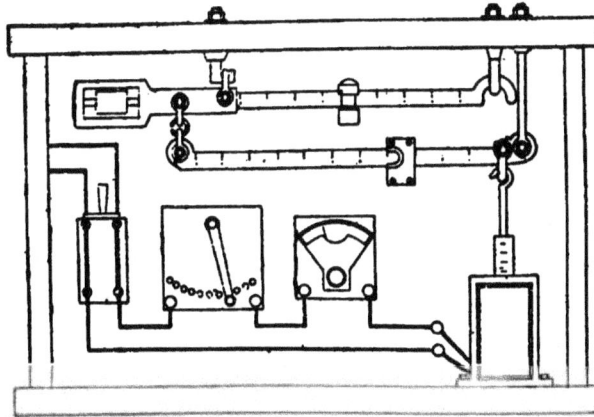

Fɪɢ. 47. — Testing Apparatus.

For the purpose of determining the actual pulls due
to varying degrees of excitation of the solenoids, the
apparatus illustrated in Fig. 47 was employed. The
plunger was attached to the scales and the counter-
weight adjusted so that the weight of the plunger was

* *Electrical World and Engineer*, Vol. XLV, 1905, pp. 796–799.

counterbalanced, and the scales balanced at zero. Hence, the weight of the plunger was entirely eliminated.

By means of an adjustable rheostat and an ammeter, any desired current strength was readily obtainable, and since the turns of each solenoid were known, the ampere-turns were easily determined.

The plunger used in this particular test was the same one used in the tests of the rim and disk solenoids and consisted of a Swedish iron bar 1 m. long and 2.87 cm. in diameter, making its cross-sectional area 6.45 sq. cm.

The reason for using coils and plungers of miscellaneous dimensions was due to the fact that these were stock sizes, but it will be seen that the lengths of the solenoids bear a constant relation to one another, and all have the same average radius.

Figure 48 shows the result of such a test. The curve marked $L = 2.54, r_a = 3$, is the smaller of the disk windings shown in Fig. 34, p. 56. It will be observed that the relation between the pull and ampere-turns for the shorter solenoid is not a straight-line proportion until approximately 6000 ampere-turns have been developed in the winding.

It will be seen that by drawing straight lines from the origin, and parallel to the curves in Fig. 48, the straight lines would represent the ampere-turns required to produce the pulls indicated, if the plunger was already saturated.

Referring to Fig. 37, p. 59, the pull due to the $L = 2.54$, $r_a = 3$ solenoid on the separately magnetized plunger is 12.2 kg. for 6000 ampere-turns. Now, in Fig. 48, the ampere-turns required for the same pull are ap-

proximately 9000. Hence, it is evident that 9000 —
6000 = 3000 ampere-turns are expended in keeping the

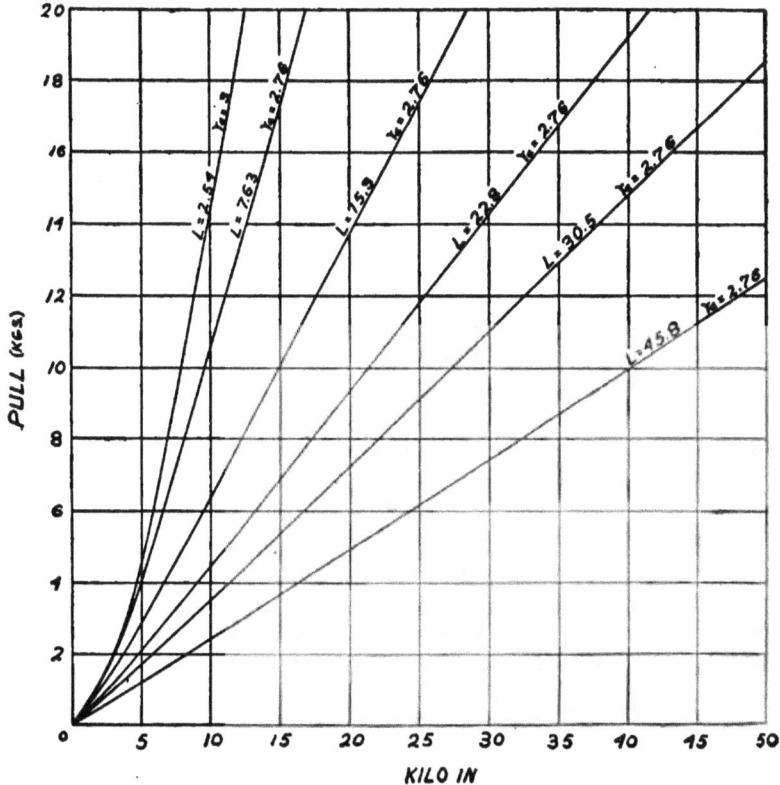

FIG. 48. — Maximum Pulls due to Practical Solenoids of Various
Dimensions.

plunger saturated. An examination of the other curves
in Fig. 48 shows similar losses, though less marked as
the length increases.

In these tests the maximum pull was taken in each
case, and since the position of maximum pull inside the
solenoid changes its position with the induction in the
plunger, it will be seen that the curves in Fig. 48 do
not necessarily represent the pulls at the exact center
of the solenoid.

This will be understood by an examination of Figs. 49 and 50 which show the pulls corresponding to differ-

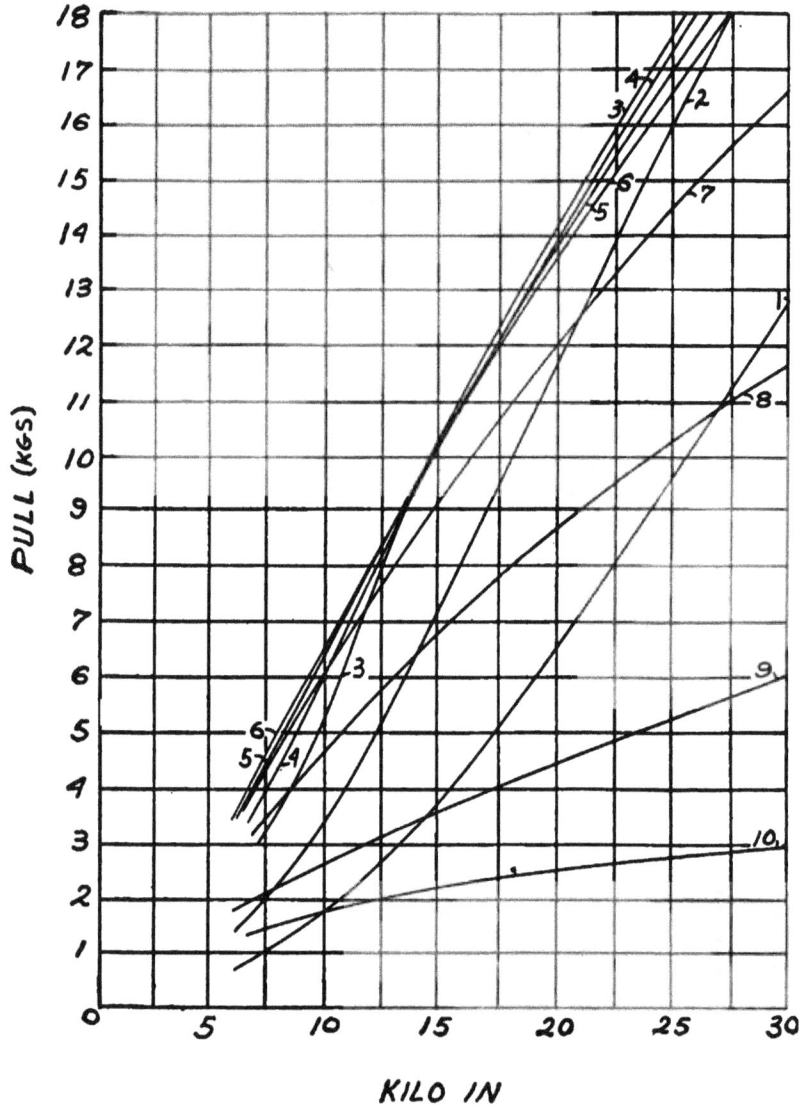

Fig. 49. — Effect of Varying Position of Plunger in Solenoid.

ent degrees of excitation expressed as kilo ampere-turns (thousands ampere-turns). These are the $L = 15.3$,

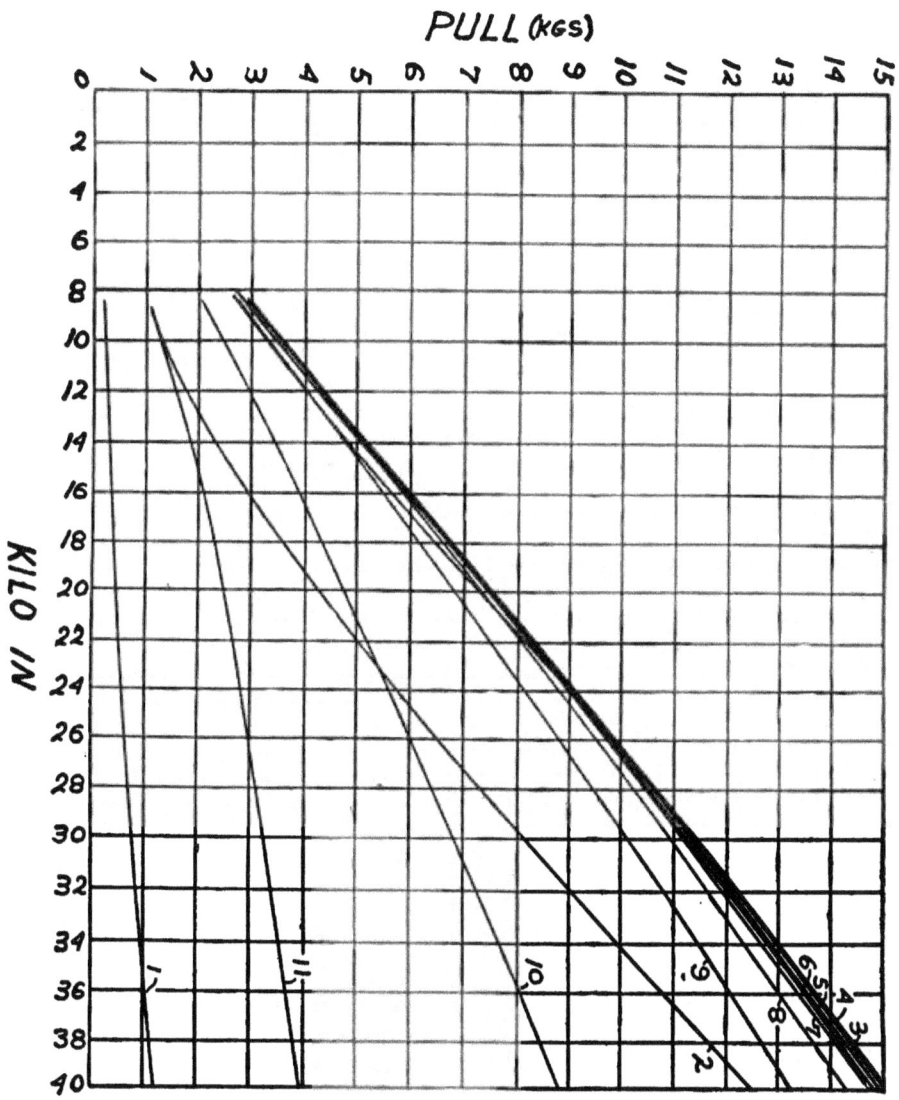

PULL (KGS)

KILO IN

Fig. 50. — Effect due to Varying Position of Plunger in Solenoid.

$r_a = 2.76$, and $L = 30.5$, $r_a = 2.76$, solenoids referred to in Fig. 48.

In Fig. 49 the curves are for the following positions of the end of the plunger:

CURVE	POSITION OF PLUNGER IN COIL
1	$\frac{1}{6}$
2	$\frac{1}{3}$
3	$\frac{5}{12}$
4	$\frac{1}{2}$
5	$\frac{7}{12}$
6	$\frac{2}{3}$
7	$\frac{3}{4}$
8	$\frac{5}{6}$
9	even with farther end
10	projecting 2.5 cm.

In Fig. 50 the positions are:

CURVE	POSITION OF PLUNGER IN COIL
1	even with end
2	$\frac{1}{6}$
3	$\frac{1}{3}$
4	$\frac{5}{12}$
5	$\frac{1}{2}$
6	$\frac{7}{12}$
7	$\frac{2}{3}$
8	$\frac{3}{4}$
9	$\frac{5}{6}$
10	even with farther end
11	projecting 2.5 cm.

It will be observed that the curves representing the relation between pull and ampere-turns before the plunger reaches the center of the winding have characteristics similar to the curves in Fig. 48, *i.e.* the straight portions of the curves do not point to the origin, but to points representing the ampere-turns required to saturate the core. These tests were also made with the plunger of 6.45 sq. cm. area, and 1 m. in length.

36. CALCULATION OF MAXIMUM PULL DUE TO SOLENOID

It has been stated that a solenoid will attract its plunger within itself until, if the plunger be the same length as the solenoid, the ends of the plunger will be even with the ends of the winding. Some characteristics will be shown presently. If now another plunger, exactly similar to the first, be held near one of the ends of the first plunger, end to end, it will be attracted to the first plunger, and then the two will be drawn inside the winding until the outer ends are equidistant from the ends of the coil, barring friction, of course. This shows plainly that a plunger longer than the winding increases the range of, and, consequently, the work due to, a solenoid.

It has been found mathematically, and confirmed by experiment, that the force required to separate the two plungers, in a long solenoid, under these conditions, that is, when the abutting ends are exactly at the center of the winding, and are perfectly joined magnetically, is

$$P_d = \frac{\mathcal{B}^2 A}{8\pi}, \qquad (66)$$

wherein P_d is the pull in dynes and $\mathcal{B}$ is the magnetic induction in the plunger.

Since 1 gram $= 981$ dynes, the pull in grams may be expressed

$$P_g = \frac{\mathscr{B}^2 A}{8\,\pi \times 981}. \tag{67}$$

In this book the unit of pull will be the kilogram (1000 grams). Representing this by P,

$$P = \frac{\mathscr{B}^2 A}{8\,\pi \times 981,000}. \tag{68}$$

From (66) the pull in dynes per square centimeter is

$$P_a = \frac{\mathscr{B}^2}{8\,\pi}. \tag{69}$$

But * $$\mathscr{B} = 4\,\pi\,\mathscr{T} + \mathscr{H}, \tag{70}$$

wherein $\mathscr{T}$ is the intensity of magnetization in the iron.

Since $$4\,\pi\mathscr{T} = \frac{\phi_i}{A} \text{ and } \mathscr{H} = \frac{\phi_a}{A},$$

wherein ϕ_i may be considered as the induced flux, ϕ_a the inducing flux, and A is the cross-sectional area of the core in square centimeters, equation (70) may be written

$$\mathscr{B} = \frac{\phi_i + \phi_a}{A}. \tag{70 a}$$

Then (69) becomes

$$P_a = \frac{1}{8\,\pi}\left(\frac{\phi_i + \phi_a}{A}\right)^2. \tag{71}$$

* *Electrical World*, Vol. LIX, No. 5, 1912, p. 1388.

Expanding (71),

$$P_a = \frac{\phi_i^2}{8\,\pi A^2} + 2\left(\frac{\phi_i\phi_a}{8\,\pi A^2}\right) + \frac{\phi_a^2}{8\,\pi A^2}. \qquad (72)$$

The pull is thus resolved into three components. Conventionally stated, $\dfrac{\phi_i^2}{8\,\pi A^2}$ is the pull between the two half-cores in Fig. 51, $2\left(\dfrac{\phi_i\phi_a}{8\,\pi A^2}\right)$ the total pull between the half-coils and the complementary half-cores, and $\dfrac{\phi_a^2}{8\,\pi A^2}$ the pull between the two half-coils.

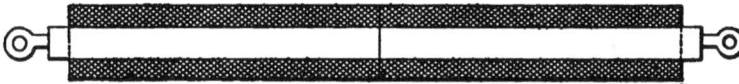

Fig. 51. — Theoretical Electromagnet.

Since, in the common types of electromagnet, the coil is not divided but is solidly wound, the latter pull is not available, so that the total pull required to separate the plunger from the stop or its complementary half-core is

$$P_a = \frac{\phi_i^2 + 2\,\phi_i\phi_a}{8\,\pi A^2}. \qquad (73)$$

When, however, the two half-cores are separated, so as to introduce an air-gap, part of the flux passes through the air-gap and part leaks from the inner ends of the half-cores to their opposite ends, so that there are now two active magnetic circuits in shunt with one another. This leakage flux is very important and should always be considered. The sum of the leakage and air-gap fluxes is nearly constant since, when the air-gap is reduced to zero, there is no leakage, and when only one core is in the coil the leakage flux is the total flux. This is illustrated in Fig. 51 a,* wherein 0.1 unit additional reluctance is assumed for the ferric parts and the fluxes are expressed as percentages.

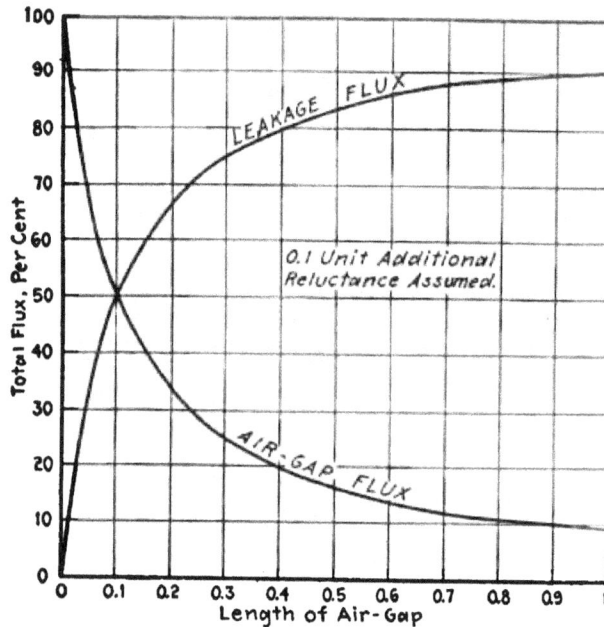

FIG. 51 a. — Relative value of fluxes.

It is therefore evident that the intensity of magnetization in the plunger or plungers persists regardless of the introduction of an air-gap between them.

In any case, the solenoid pull is proportional to $2\,\mathcal{H}(\mathcal{B}-\mathcal{H})$.

Now
$$\mathcal{H} = \frac{0.2\,\pi IN\theta_t}{L}. \qquad (64)$$

* *Electrical World*, Vol. LXIII, No. 5, 1914, p. 260.

In this book, the value of $\mathcal{B} - \mathcal{H} = 20,000$ will be assumed in all future calculations of the solenoid for thoroughly saturated plungers of soft iron or steel.

Then
$$P = \frac{20,000 \times 0.4\,\pi I N \theta_t A}{8\,\pi \times 981,000\,L},$$

or
$$P = \frac{I N \theta_t A}{981\,L}. \tag{74}$$

37. Ampere-turns Required to Saturate Plunger

In (74) the losses due to the ampere-turns required to keep the plunger saturated have not been considered. These must be allowed for in solenoids up to 25 or 30 cm. in length, but for greater lengths, the losses are inappreciable.

Assigning λ to this loss, (74) becomes
$$P = \frac{A\theta_t(IN - A\lambda)}{981\,L}, \tag{75}$$

as the loss will vary with the cross-sectional area of the plunger for the same ampere-turns.

In Fig. 52 are shown the approximate values of λ as taken by observation from tests.

From (75)
$$IN = \frac{981\,PL}{A\theta_t} + A\lambda, \tag{76}$$

and since
$$\theta_t = 2 - \frac{r_a}{1.07\,L} \text{ (approx.)}, \tag{63}$$

(76) may be written
$$IN = \frac{981\,PL}{A\left(2 - \dfrac{r_a}{1.07\,L}\right)} + A\lambda. \tag{77}$$

This formula will be found quite accurate, but it is well to increase the calculated ampere-turns, to allow

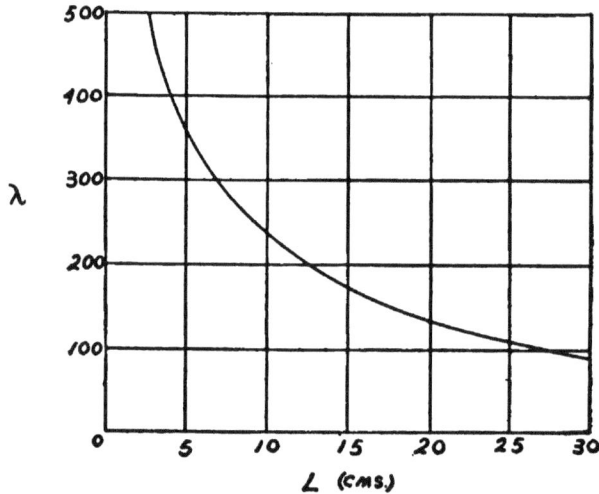

FIG. 52. — Approximate Ampere-turns required to saturate Plunger.

for variation in the value of $\mathscr{B}$ in the plunger, and since the ampere-turns decrease with a rise in temperature of the winding of the solenoid, for a given e. m. f., it is always better to have a little too much pull than not quite enough.

The weight of the plunger, as well as losses due to friction, should be allowed for, according to the conditions under which the solenoid is to be operated, since in the formulæ the weight of the plunger is not considered, and in the tests referred to, the plunger was counterbalanced.

Substituting the value of θ_t from (63) in (74),

$$P = \frac{A(IN - A\lambda)\left(2 - \dfrac{r_a}{1.07\,L}\right)}{981\,L}. \qquad (78)$$

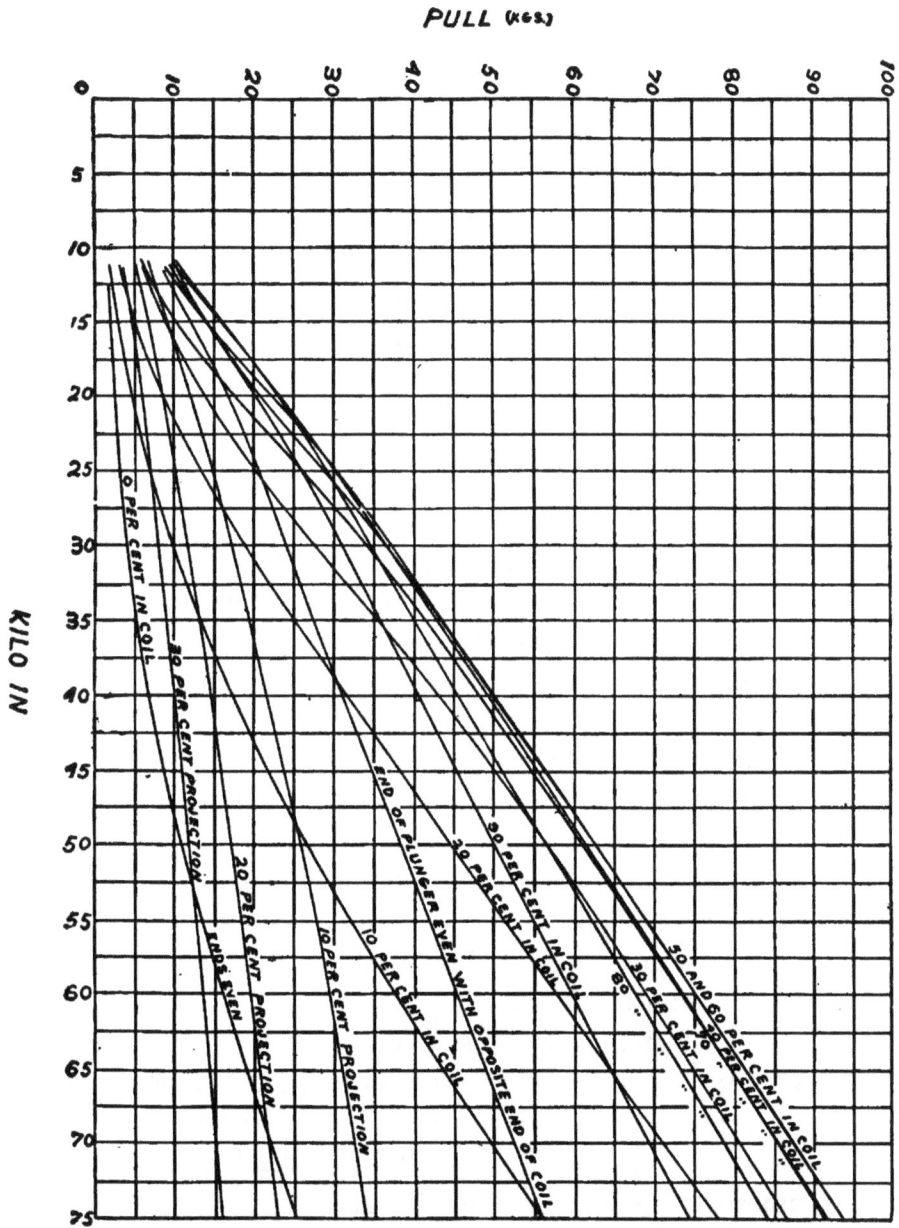

FIG. 53. — Characteristic Force Curves of Solenoid.

Figure 53 is the result of a test of a solenoid of the following dimensions: $L = 25.4$, $r_a = 6.8$, $T = 8.6$. The plunger was of soft steel 17.9 sq. cm. in area and 60 cm. long. In this as was the case with all the solenoids tested (the rim solenoids excepted), the internal diameter of the solenoid was made as small as possible; the brass tube, insulation, and sufficient freedom for the plunger being the controlling factors.

The values in Fig. 53 compare very favorably with values calculated by the above formulæ.

38. RELATION BETWEEN DIMENSIONS OF COIL AND PLUNGER

The general construction of the solenoid may vary with the ideas of the designer, but the dimensions of the winding and plunger are all important, and it must be remembered that the strength of a solenoid is limited by the carrying capacity of the winding.

As the total work obtained from any long solenoid is practically constant, regardless of its actual length, for the same ampere-turns, the pull for any long solenoid is proportional to the ampere-turns per unit length. Therefore, a 50-cm. solenoid will have practically the same pull as a 25-cm. solenoid, if twice the energy is applied to the winding.

The diameter of the solenoid will vary with the diameter of the core or plunger, and other conditions, such as heating, etc., but a good general rule is to assume a diameter for the solenoid equal to about three times the diameter of the plunger.

An examination of Fig. 48 will show that the ampere-turns per square centimeter of core should never fall below 1000 for the shorter solenoids, in order to keep the core well saturated. Furthermore, it is desirable to work the cores of solenoids at high densities because of the relative weight of the plunger when working with lower densities, and the pull per ampere-turn is much less at low densities. If the core was not saturated, the pull would not be directly proportional to the ampere-turns, as the permeability of the iron would be very changeable at low densities.

On the other hand, it is not economical to have more ampere-turns than are necessary to saturate the plunger. Hence, the following method may be adopted.

Let INA = the product of the minimum ampere-turns and the minimum cross-sectional area of the plunger necessary to keep the plunger saturated and produce the required pull, IN_c = the minimum ampere-turns per square centimeter of plunger when the latter is saturated, and A = cross-sectional area of the plunger in square centimeters. Then,

$$A = \sqrt{\frac{INA}{IN_c}}, \qquad (79)$$

and $INA = IN_c A^2$ (80). This will be understood by referring to Fig. 54.

As an example, assume that $IN_c = 1000$, which is a good average value to use in practice. Then $INA = 1000 A^2$, whence

$$A = \sqrt{\frac{INA}{1000}} = \frac{\sqrt{INA}}{31.6}.$$

By this method the ampere-turns will always be 1000 for each square centimeter of plunger, which insures the minimum expenditure in watts to produce a given result.

As before stated, other values than $IN_c = 1000$ may be chosen, but as the solenoids in common use are not

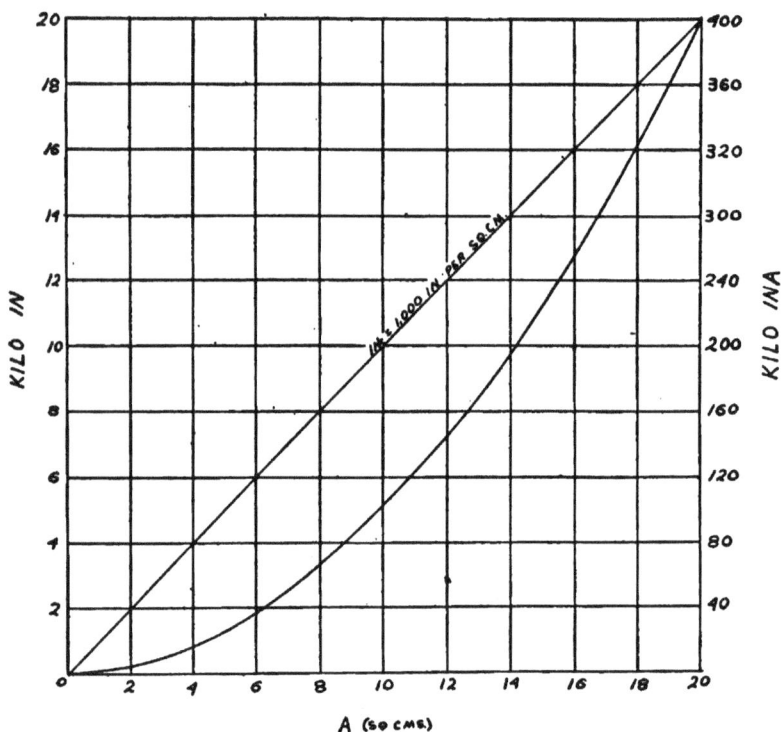

Fig. 54. — Ratio between Ampere-turns and Cross-sectional Area of Plunger.

very great in length, the author has found the above value to be approximately correct for general practice, although $IN_c = 1250$ is a safer value.

Since $A = d_p{}^2 \times 0.7854$ (81), the diameter of the plunger d_p may be calculated direct by substituting the value of A from (81) in (80). Then,

$$INA = 0.617\ IN_c d_p{}^4,\qquad(82)$$

and

$$d_p = \sqrt[4]{\frac{INA}{0.617\ IN_c}}.\qquad(83)$$

$$IN = \frac{INA}{A} = IN_cA. \tag{84}$$

Hence,
$$IN = \frac{0.617\, IN_c d_p^4}{A}, \tag{85}$$

or
$$IN = \frac{0.617\, IN_c d_p^4}{0.7854\, d_p^2} = 0.7854\, IN_c d_p^2, \tag{86}$$

and
$$d_p = \sqrt{\frac{IN}{0.7854\, IN_c}}, \tag{87}$$

which means, of course, that the ampere-turns will be IN_c times the cross-sectional area of the plunger.

For short solenoids, $IN - A\lambda$ must be substituted for IN, but, in general, the above method is near enough.

39. RELATION OF PULL TO POSITION OF PLUNGER IN SOLENOID

It was stated that when the end of the plunger enters the solenoid, the pull increases until the position of maximum pull is reached, when it again falls off until, if the plunger be the same length as the solenoid, the pull will fall to zero when the ends of the former are even with those of the latter, while if the plunger be longer than the solenoid, the end of the plunger will protrude a short distance from the solenoid.

Figures 55 to 58 show the characteristics of the solenoids of constant radius ($r_a = 2.76$) and lengths 15.3, 22.8, 30.5, and 45.8 cm., respectively, with the plunger 1 m. in length, and 6.45 sq. cm. in cross-section.

It will be observed that as the length increases, the pull is more uniform over a given distance, and that by assuming a pull lower than the actual pull, the range may be somewhat increased.

PULL (KGS.)

IN = 15,000

PLUNGER IN COIL (CMS.)

FIG. 55. — Characteristics of
Solenoid 15.3 cms. long.

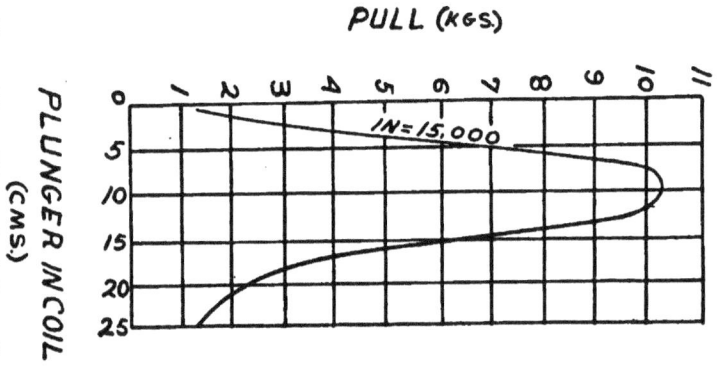

PULL (KGS.)

IN = 18,200

IN = 41,000

PLUNGER IN COIL (CMS.)

FIG. 58. — Characteristics of Solenoid 45.8 cms. long.

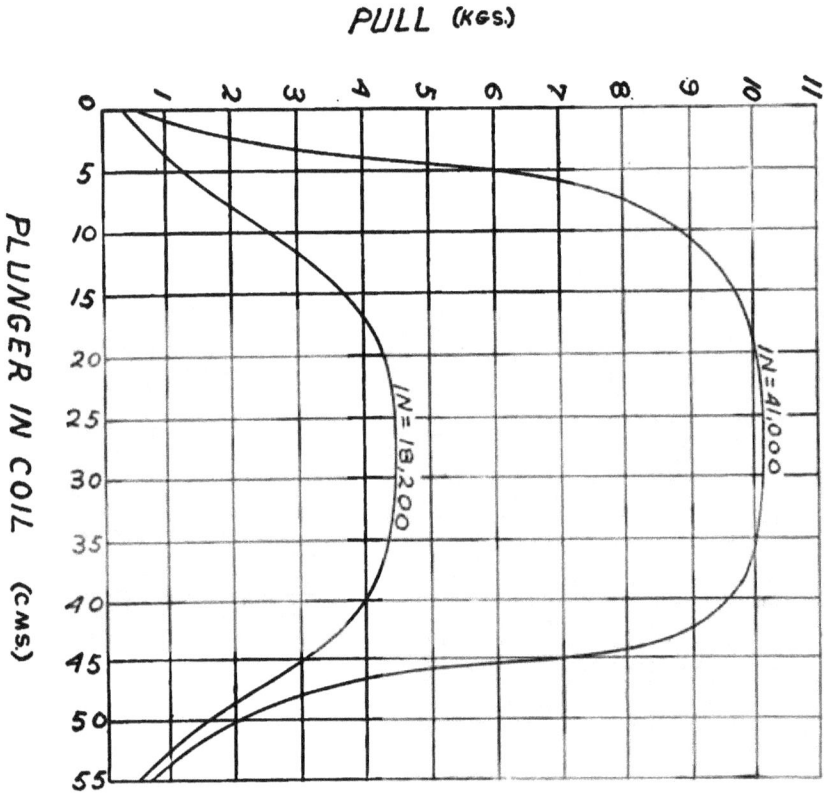

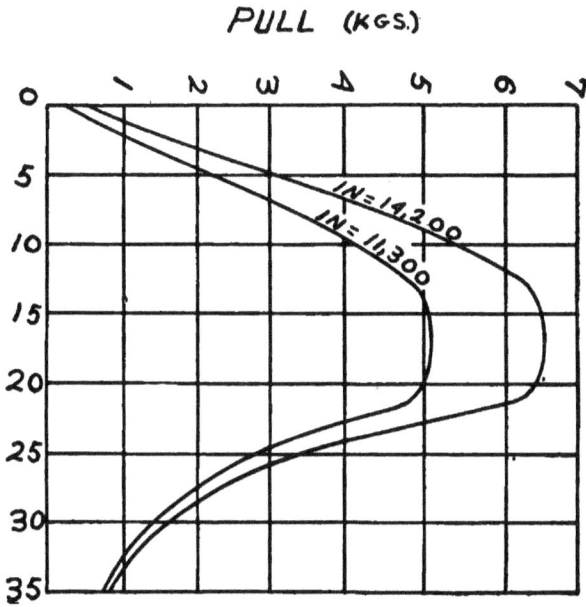

PULL (KGS.)

IN = 14,200
IN = 11,300

PLUNGER IN COIL (CMS.)

FIG. 56.—Characteristics of Solenoid 22.8 cms. long.

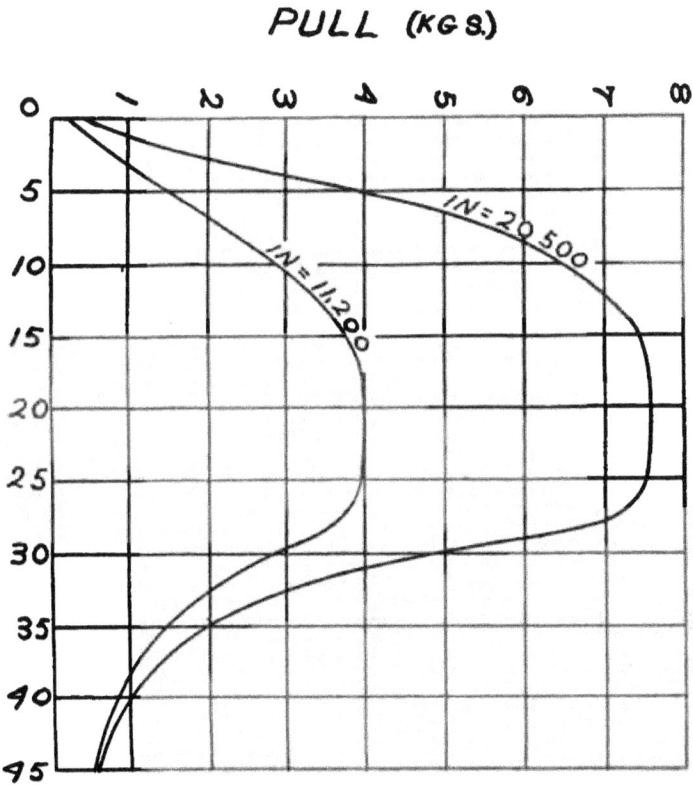

PULL (KGS.)

IN = 20,500
IN = 11,200

PLUNGER IN COIL (CMS.)

FIG. 57.—Characteristics of Solenoid 30.5 cms. long.

In Fig. 59 is shown the characteristics of the 45.8-cm. solenoid with a plunger of the same length as the

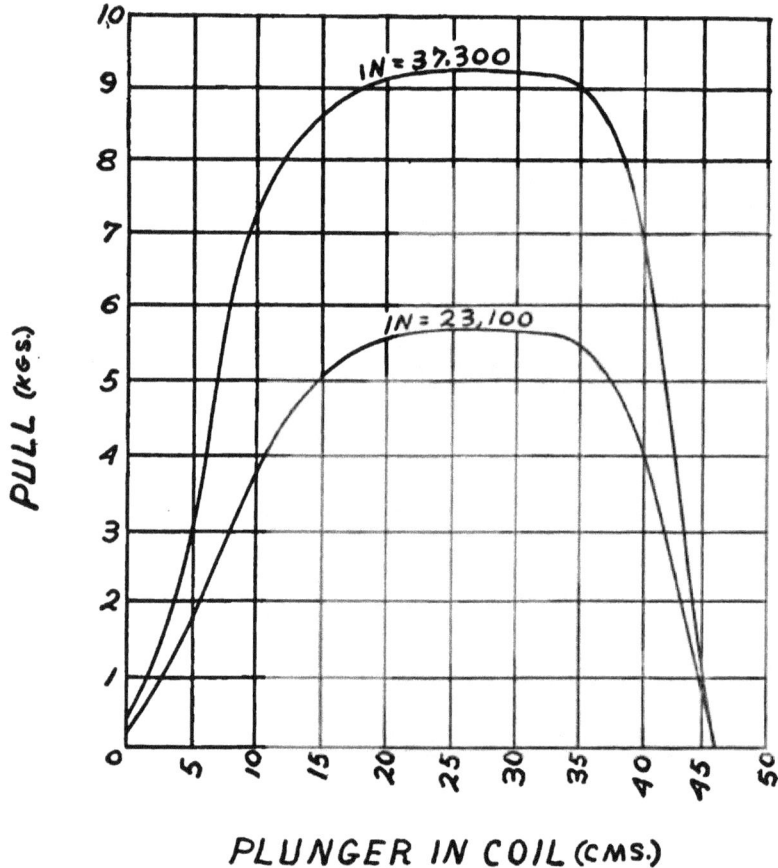

FIG. 59.—Characteristics of 45.8-cm. Solenoid with Plunger of Same Length.

solenoid. It will be noticed that a greater range of action may be obtained with a plunger longer than the solenoid, and that the range increases with the ampere-turns. This latter effect is due to the fact that the ratio of total ampere-turns to those required to saturate the plunger is greater for large values of IN.

The pull, however, will not be so great where the

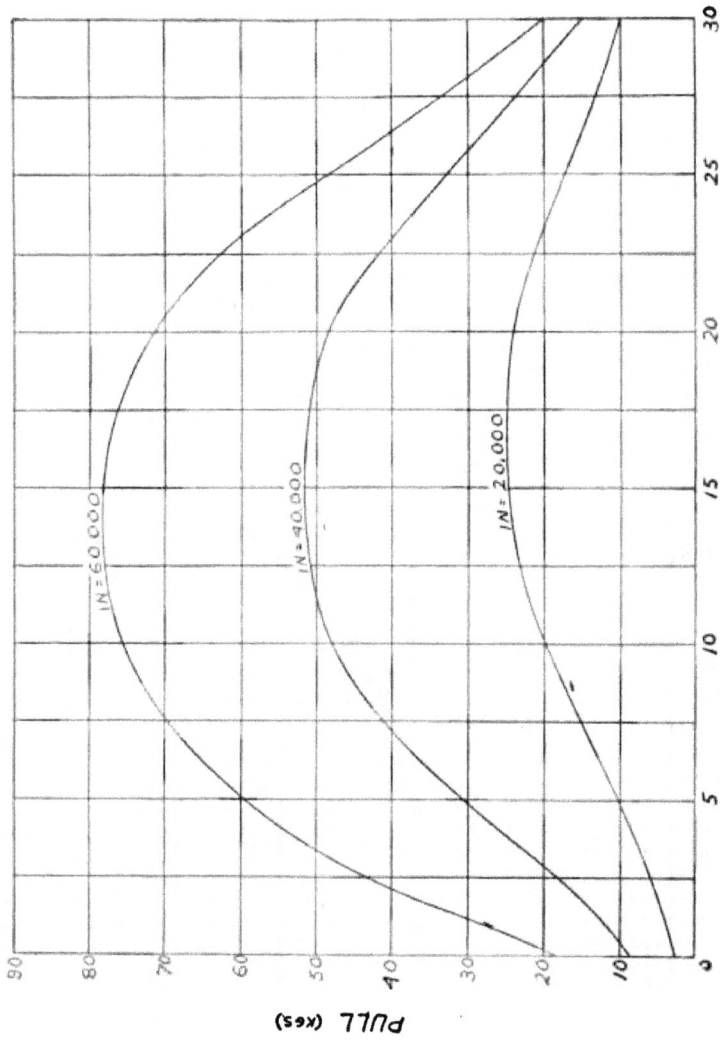

PLUNGER IN COIL (CMS.)

FIG. 60. — Characteristics of Solenoid 25.4 cms. long.

plungers are the same length as the coils, excepting in cases of comparatively long solenoids.

In Fig. 60 are shown the force curves due to the 25.4-cm. solenoid previously referred to. This is plotted

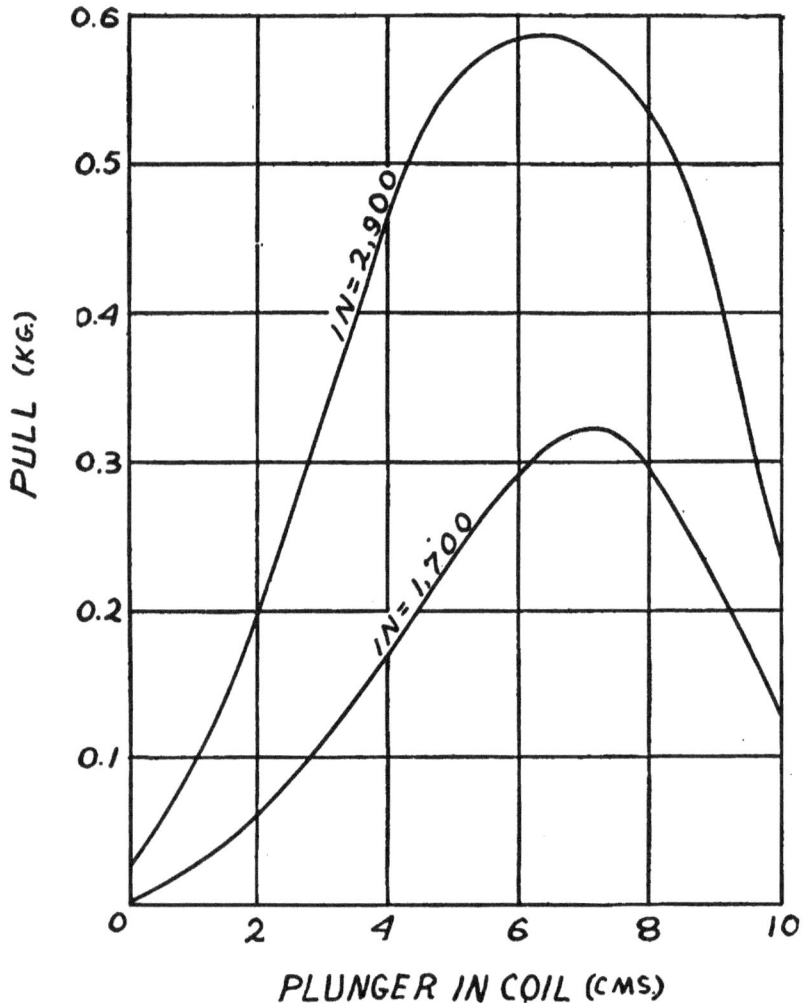

Fig. 61. — Characteristics of Solenoid 8 cms. long.

from Fig. 53. The curves in Fig. 61 are due to a solenoid 8 cm. long, with a soft steel plunger 1 sq. cm. in cross-section, and 30 cm. long. The other dimensions are $T = 2$, $r_a = 1.09$.

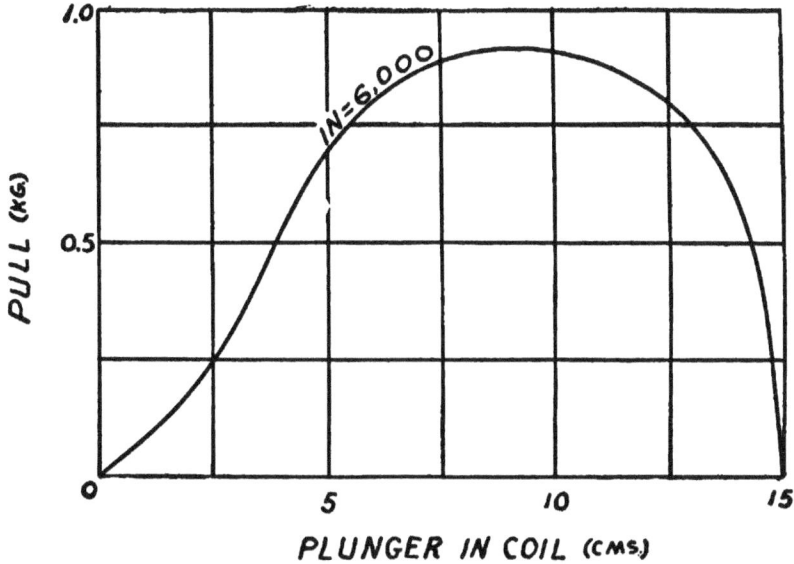

FIG. 62.—Characteristics of Solenoid 15 cms. long.

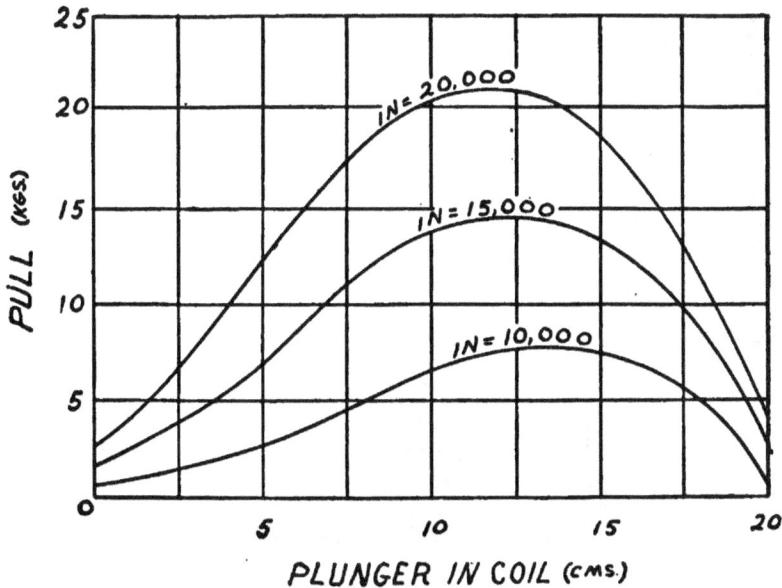

FIG. 63.—Characteristics of Solenoid 17.8 cms. long.

The characteristics due to a solenoid in which $L = 15$, $T = 1.17$, and $r_a = 0.66$, are shown in Fig. 62. The plunger was 1.27 sq. cm. in cross-section.

Figure 63 is due to a solenoid of dimensions $L = 17.8$, $T = 2.54$, $r_a = 3.5$. The sectional area of the plunger was 11.4 sq. cm. and the length 25.4 cm.

The above cases are mentioned at random in order to show the general relations existing between solenoids of all dimensions. The maximum pulls for any of these solenoids may be closely approximated by use of formula (78).

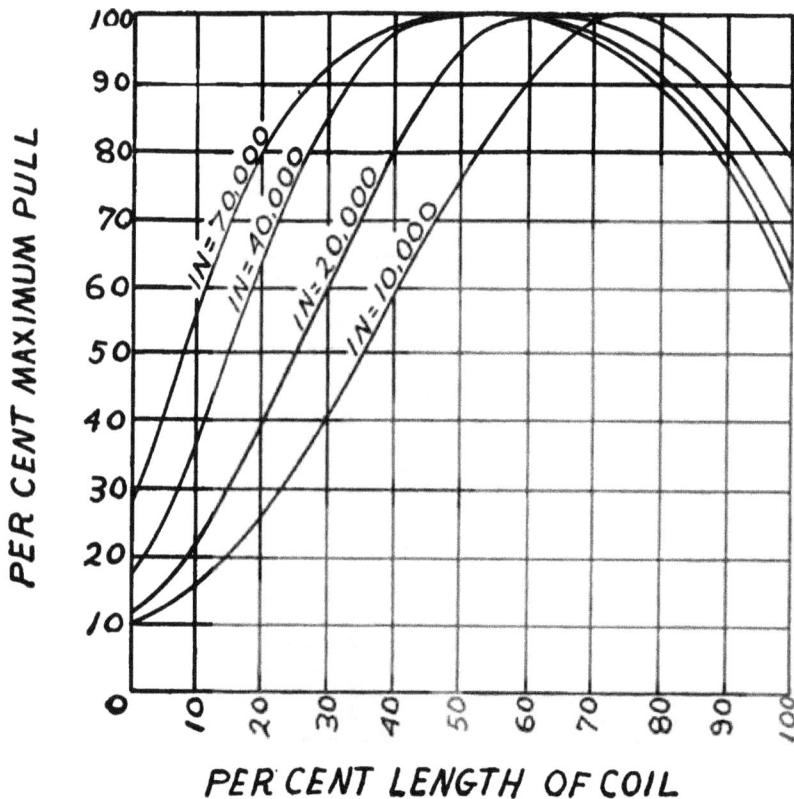

FIG. 64. — Effect of increased m. m. f. on Range of Solenoid.

40. CALCULATION OF THE PULL CURVE

A comparison of the curves in Figs. 55 to 63 with the theoretical force curve in Fig. 28 will show that, while there is a resemblance, the position of maximum pull is shifted toward the farther end of the solenoid for very weak magnetizing forces, while the theoretical force curve is the same at both sides of the vertical center of the solenoid.

This is due to the fact that for magnetizing forces of low value the plunger may not become saturated until

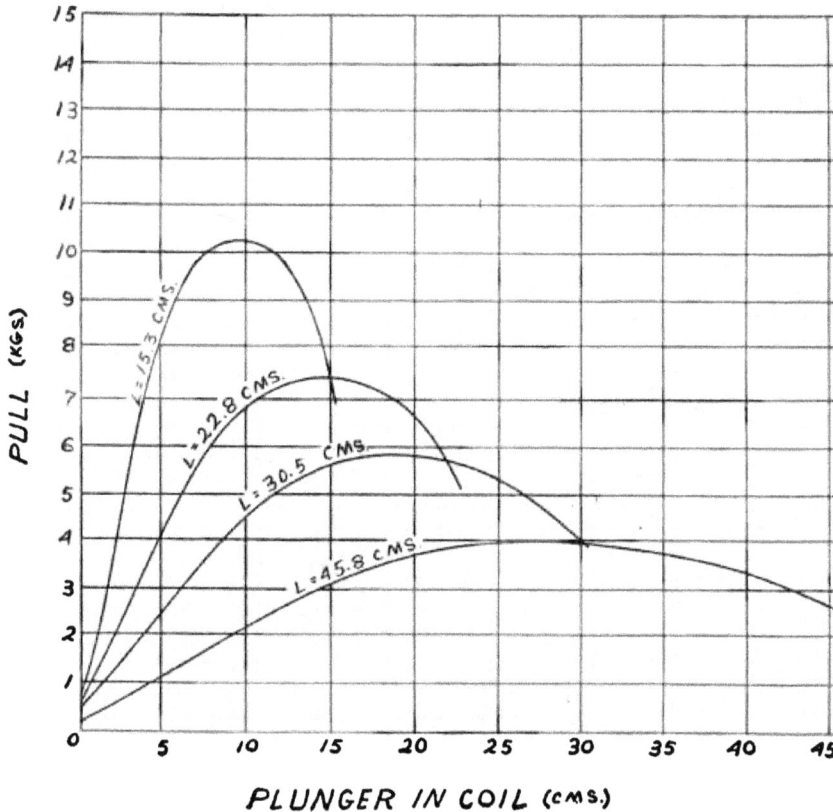

FIG. 65.—Comparison of Solenoids of Constant Radii, but of Different Lengths.

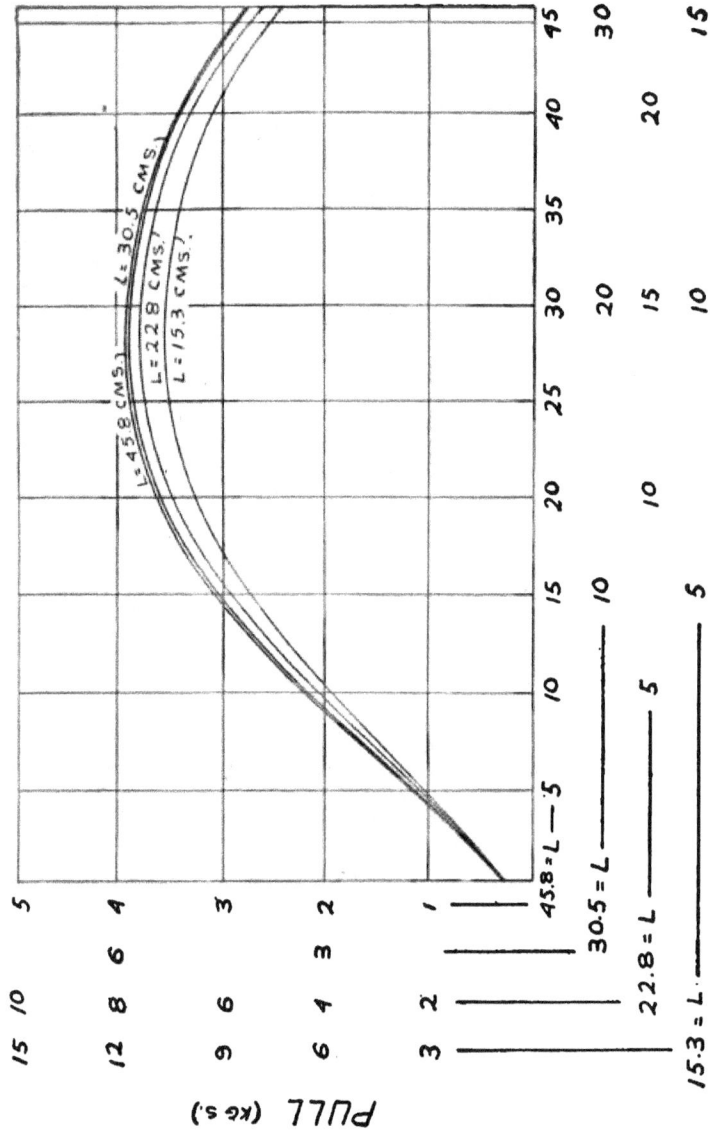

PLUNGER IN COIL (CMS.)

Fig. 66.—Curves in Fig. 65 reduced to a Common Scale.

it has protruded a short distance from the farther end of the solenoid. Figure 64, in particular, shows this effect, which is plotted from Fig. 53. In these, however, the plunger has become saturated before it has reached the farther end of the solenoid, owing to sufficient magnetizing force to saturate the plunger.

It can be shown that, in general, the curves of all solenoids are very similar for equal degrees of magnetization.

In Fig. 65 are shown curves due to the solenoids of constant radius ($r_a = 2.76$) and of lengths 15.3, 22.8, 30.5, and 45.8 cm., respectively, tested with the plunger one meter in length, and with 15,000 ampere-turns. The plunger was thoroughly saturated, since the ampere-turns per square centimeter of core were 2300.

By grouping the curves in Fig. 65 on a common plane, so as to make the ampere-turns per unit length the same in all cases (see Fig. 66), it is seen that the curves are similar, with the exception that the peaks are proportionately higher for the longer solenoids. If, however, the pulls throughout the entire range of each curve in Fig. 66 are compared with the maximum pull for that solenoid, the curves are found to be practically similar in all cases cited. This common curve is illustrated in Fig. 67.

It, therefore, remains to determine the equation for this curve, which is partly sinusoidal, and the equation $y = \sin 0.77\, x$ (88) satisfies this condition for practical purposes, as reference to Fig. 68 will show. In this case the length of the solenoid is compared with 180 degrees.

While the solenoid pull curve in Fig. 68 is slightly higher than the $y = \sin 0.77\, x$ curve from 25 degrees to

120 degrees, it must be understood that the percentage
of maximum pull throughout the first half of the sole-

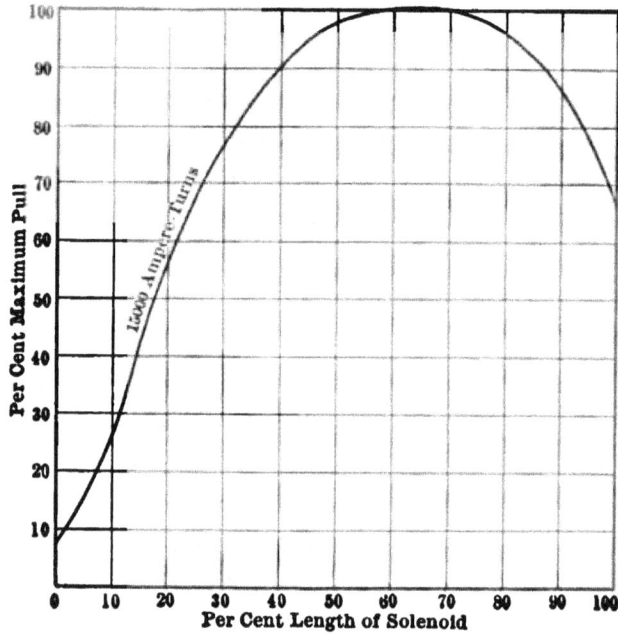

Fig. 67. — Average of Curves.

noid is greater for higher than for lower magnetizing
forces, owing to the fact that the plunger is more quickly

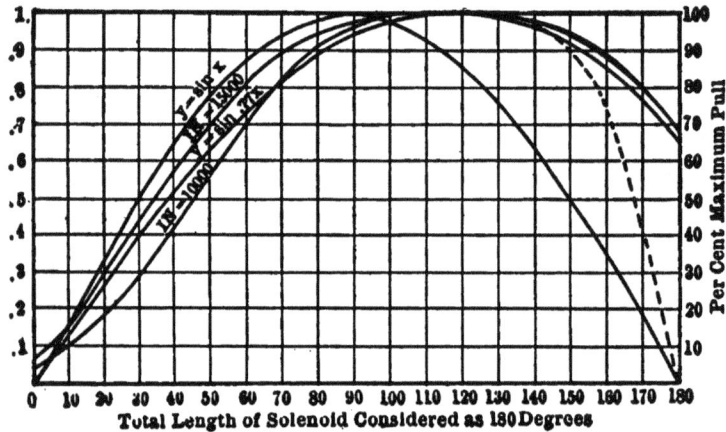

Fig. 68. — Average Solenoid Curve compared with Sinusoid.

saturated under the former condition, thereby increasing the pull — or, to be exact, the percentage of maximum pull; and since this curve represents the pull due to 15,000 ampere-turns, the percentage of maximum pull would be somewhat lower with a magnetizing force just sufficient to saturate the plunger at the position of maximum pull, and, therefore, the curve $y = \sin 0.77\,x$ represents a good average, as the curves beyond the position of maximum pull do not vary appreciably as the magnetizing force increases, after the plunger is saturated.

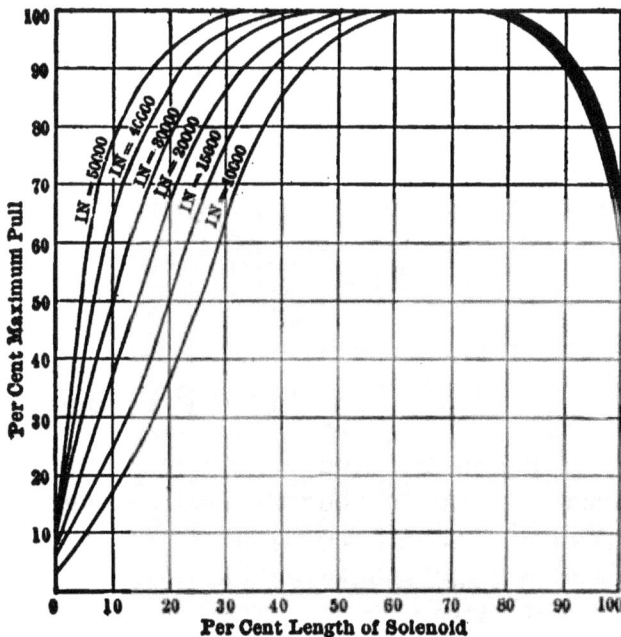

FIG. 69. — Effect of Increasing Ampere-turns.

This effect is illustrated in Fig. 69, which is the result of a test of the 30.5-cm. solenoid. An inspection of Fig. 69 also shows that the range of the solenoid is much greater with high than with low magnetizing forces.

Again referring to Fig. 68, the dotted curve represents the effect of using a plunger of the same length as the solenoid, in the case of a long solenoid.

Now from these data a curve may be plotted showing the approximate pull at all points throughout the range of the solenoid. It will be recalled that the ratio of the actual pull to the maximum (y) is equal to sin $0.77\,x$; and if we let l_2 represent the distance in centimeters, the plunger is in the coil, and compare the linear ratios with those represented by degrees, we have,

$$x = \frac{180\,l_2}{L}. \tag{89}$$

Representing the pull at any point by p, the ratio of actual to maximum pull is

$$\frac{p}{P} = \sin\frac{0.77 \times 180\,l_2}{L},$$

whence
$$p = P\sin\frac{138.6\,l_2}{L}. \tag{90}$$

41. POINTED OR CONED PLUNGERS

By pointing or tapering the plunger, varying force curves may be obtained, the pull usually being maximum when the pointed end has protruded from the other end of the coil to which it entered, this depending upon the amount of tapering.

In this case, however, the pull is not so great as with the regular plunger, owing to the diminished cross-sectional area of the iron due to tapering.

It is obvious that almost any desired form of curve may be obtained by giving the plunger peculiar shapes.

42. Stopped Solenoids

The effective range of a solenoid may be greatly increased by placing a piece of iron at the farther end of the solenoid. This is commonly called a *Stop*, and may project within the winding if desired.

Figure 70 shows an experimental solenoid * designed by the author for tests of the *Stopped Solenoid*. The extension piece on the end is provided with a thumbscrew which permits of cores of various lengths being fastened in any desired position.

In the tests to be described, all cores were of soft iron, 1.27 sq. cm. cross-sectional area, and in each case there were 6300 ampere-turns in the winding. The dimensions of the coil were $L = 15$, $r_a = 0.66$ cm.

Curve *a* in Fig. 71 is due to the attraction between the magnetizing force in the coil and the flux in the plunger, which was 15 cm. in length. Curve *b* was obtained from the same solenoid, but with a stop 2.54 cm. long, its right-hand end being even with the right-hand end of the winding. Curve *c* is

* *American Electrician*, Vol. XVII, 1905, pp. 299–302.

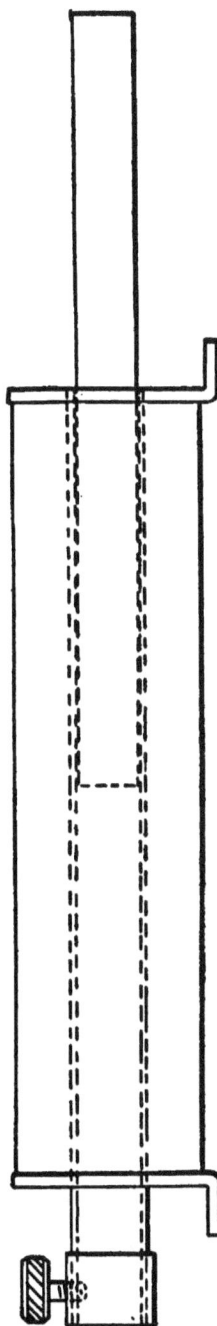

Fig. 70.—Experimental Solenoid.

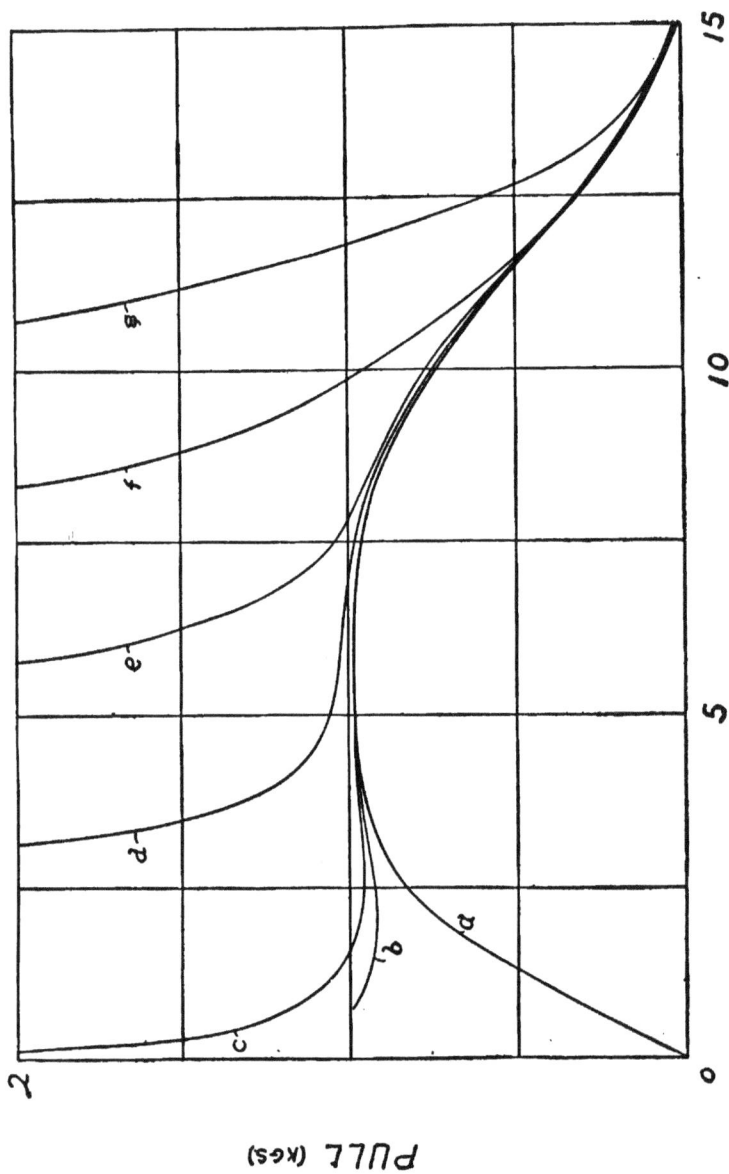

FIG. 71.—Characteristics of Experimental Solenoid.

from a test with a stop of approximately the same length as the plunger, and with its right-hand end even with the left-hand end of the winding, as in the case of curve *b*.

It is obvious that a single piece of iron at the end of the solenoid is beneficial, as it prevents the pull from falling off at that end, thereby increasing the effective range. It should be noted that all parts of the frame of this experimental solenoid were of non-magnetic material, the only ferric portions being the plunger and stop.

The curves *d, e, f,* and *g* are due to the end of the stop being inserted in the coil one sixth, one third, one half, and two thirds of the length of the solenoid, respectively.

An examination of Fig. 71 shows that the curves *c, d, e,* etc., are the result of the attraction between the plunger and the stop plus the solenoid effect.

It will be noted that stronger pulls are obtained with the long stop than with the short one. This is due to the fact that the longer rod offers a better return path for the lines of force, thereby increasing the induction and, consequently, the pull.

Thus far we have been dealing with solenoids which depend upon the air or surrounding region as a return path for the lines of force. It has been seen that by lengthening the plunger, or the stop of the stopped solenoid, a greater pull was obtained, due to the greater superficial area of the iron. It is, therefore, evident that if the pull is increased by this slight addition in iron, the pull should be very much increased, for short air-gaps within the winding, between the plunger and the stop, if the return circuit consisted wholly of iron.

The fact of placing iron around the outside of the solenoid will not, however, materially increase the pull due to the purely solenoid effect.

CHAPTER VII

IRON-CLAD SOLENOID

43. EFFECT OF IRON RETURN CIRCUIT

IN practice the solenoid is usually provided with an external return circuit of iron or steel, as in Fig. 72.

The conditions now are quite different than in the simple solenoid. In this case, no matter how much the length of the plunger may exceed that of the winding, the pull, for the same coil with a given magnetizing force, will be practically the same at points near the center of the winding, whereas in the simple solenoid the pull will vary with the length of the plunger, unless the coil be very long, owing to the greater surface of the protruding plunger, which decreases the reluctance of the return circuit outside of the winding.

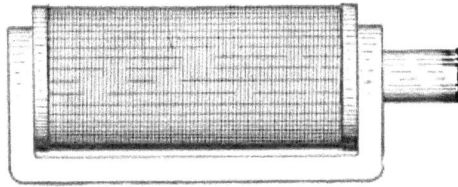

FIG. 72. — Iron-clad Solenoid.

In the case of the iron-clad solenoid the lines of force are nearly all concentrated within the center of the coil, and return through the iron frame. Hence, no appreciable attraction will occur between the coil and plunger, until the latter is inserted through the opening in the iron frame, and within the winding.

Likewise, the pull will cease when the end of the plunger within the winding touches, or protrudes into, the farther end of the iron frame, depending upon whether the frame at this end is solid or is bored through.

44. CHARACTERISTICS OF IRON-CLAD SOLENOIDS

Curve *a* in Fig. 73 is due to the 30.5-cm. simple solenoid referred to on p. 95, the plunger being 6.45 sq. cm. in cross-section, as before. Curve *b* is due to a test with the same coil, but with an iron return circuit, as in Fig. 72.

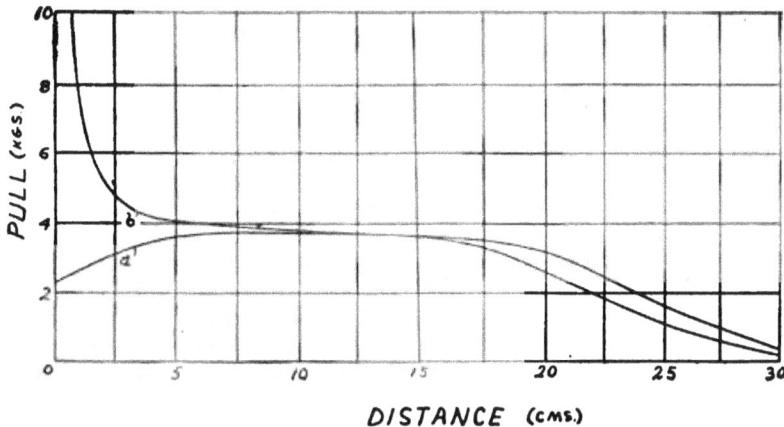

FIG. 73. — Characteristics of Simple and Iron-clad Solenoids.

It will be observed that the uniform range is considerably increased, and that the pull is very strong as the end of the plunger within the winding approaches the iron frame.

By boring a hole clear through the rear end of the iron frame, as in Fig. 74, any jar due to sudden stoppage of the plunger under

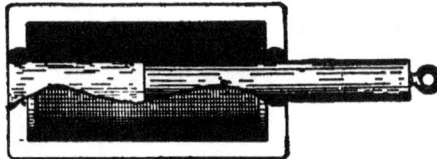

FIG. 74. — Magnetic Cushion Type of Iron-clad Solenoid.

this strong attraction may be avoided. If this strong pull at the end of the stroke is undesirable, only the uniform or slightly accelerating portion of the pull curve may be used.

45. Calculation of Pull

The pull in kilograms due to an iron-clad solenoid, at the center of the winding, is calculated by the same formula as for the simple solenoid; that is,

$$P = \frac{A\theta_t \ (IN - A\lambda)}{981 \ L}, \qquad (75)$$

wherein

$$\theta_t = 2 - \frac{r_a}{1.07 \ L} \qquad (63) \quad (\text{See p. 68.})$$

Hence,

$$IN = \frac{981 \ PL}{A\left(2 - \dfrac{r_a}{1.07 \ L}\right)} + A\lambda. \qquad (77)$$

Besides the pull due to what may be termed the pure solenoid effect, attraction takes place between the end of the plunger within the solenoid and the farther end of the iron frame. This latter effect may be approximately expressed by formula (68).

$$P = \frac{\mathscr{B}^2 A}{8 \ \pi \times 981,000}.$$

Now, since practically the entire reluctance of the magnetic circuit is in the air-core or air-gap, the reluctance of the iron frame may be neglected in the design of iron-clad solenoids.

Hence, assuming that the total reluctance is in the air-gap, (45) may be written

$$\mathscr{B} = \frac{1.25664 \ IN}{l}, \qquad (91)$$

since the permeability of air is unity.

Substituting this value for $\mathscr{B}$ in (68),

$$P = \left(\frac{1.25664 \ IN}{l}\right)^2 \times \frac{A}{8 \ \pi \times 981,000},$$

whence

$$P_m = A \left(\frac{IN}{3951 \ l}\right)^2. \qquad (92)$$

The pull is expressed in kilograms as before. P_m represents the purely magnetic pull between the plunger and the stop, and P_s is the total pull due to an iron-clad solenoid. Hence, $P_s = P + P_m$ (93), whence

$$P_s = \frac{A\theta_t (IN - A\lambda)}{981\,L} + A \left(\frac{IN}{3951\,l}\right)^2, \qquad (94)$$

or $\qquad P_s = A \left[\frac{\theta_t (IN - A\lambda)}{981\,L} + \left(\frac{IN}{3951\,l}\right)^2 \right]. \qquad (95)$

The iron or steel frame need not be very great in cross-section, unless the strong pull near the end of the stroke is to be taken advantage of.

46. EFFECTIVE RANGE

Figures 75 to 78 show the characteristics of several iron-clad solenoids. These had cast-iron frames and

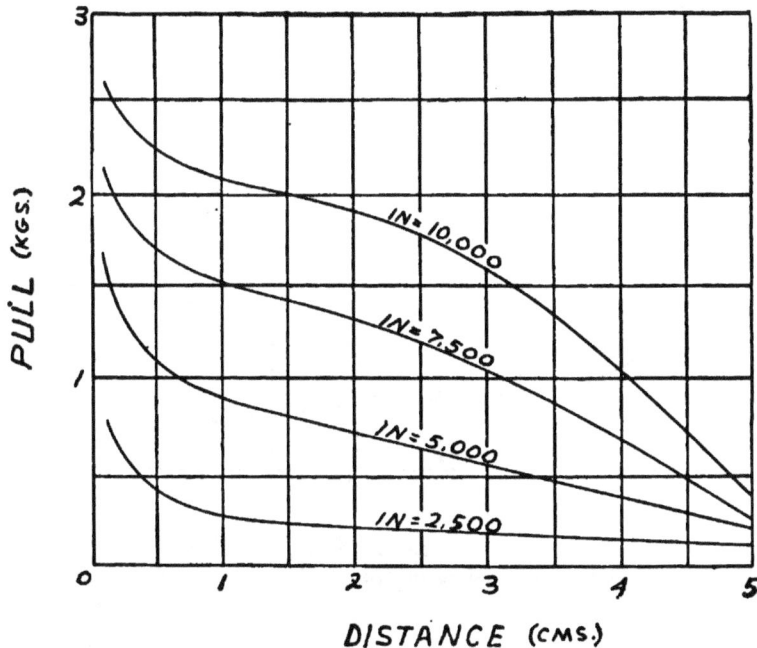

FIG. 75.—Characteristics of Iron-clad Solenoid. $L = 4.6$.

soft-iron plungers, and were of the general construction
of the iron-clad solenoid in Fig. 74, with the brass tube

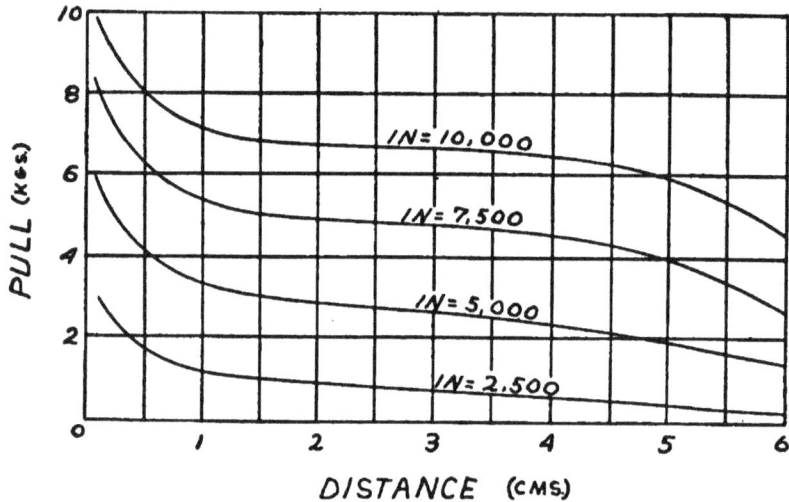

FIG. 76. — Characteristics of Iron-clad Solenoid. $L = 8.0$.

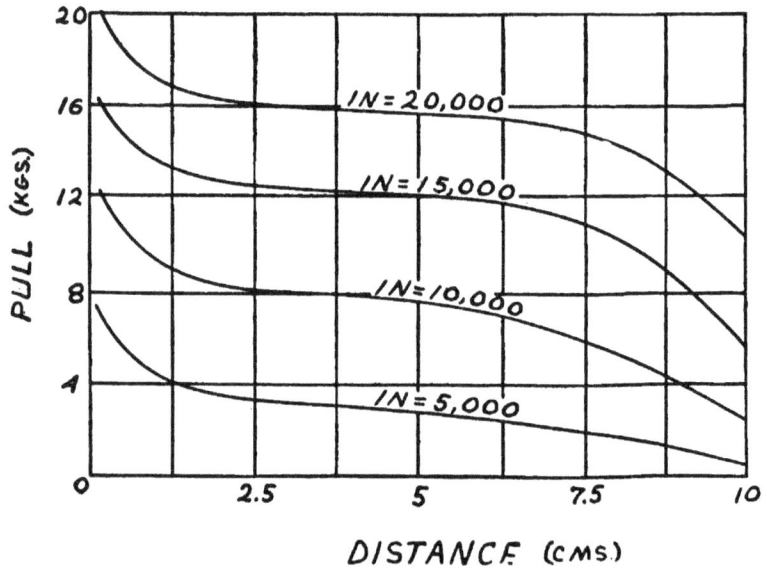

FIG. 77. — Characteristics of Iron-clad Solenoid. $L = 11.4$.

in which the plunger moves passed clear through
openings in the iron frame at both ends. This is

known as the *magnetic cushion* type, as there is no jar when the plunger completes its stroke, even when the

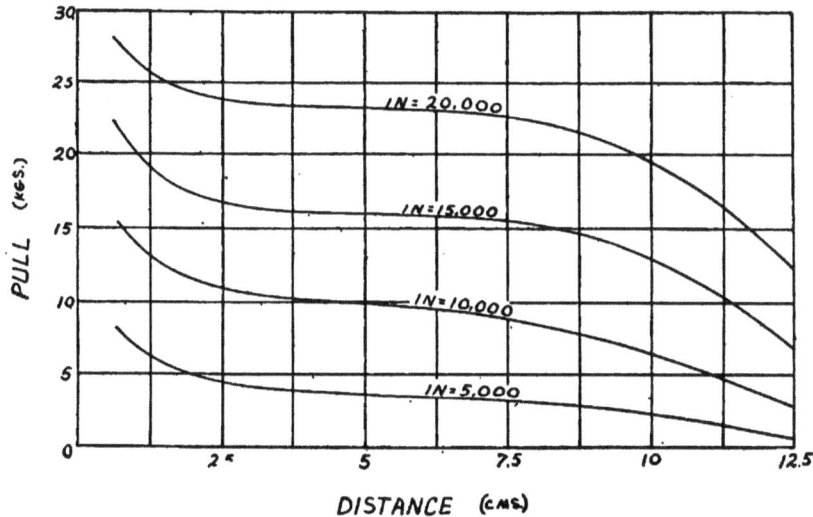

FIG. 78. — Characteristics of Iron-clad Solenoid. $L = 15.2$.

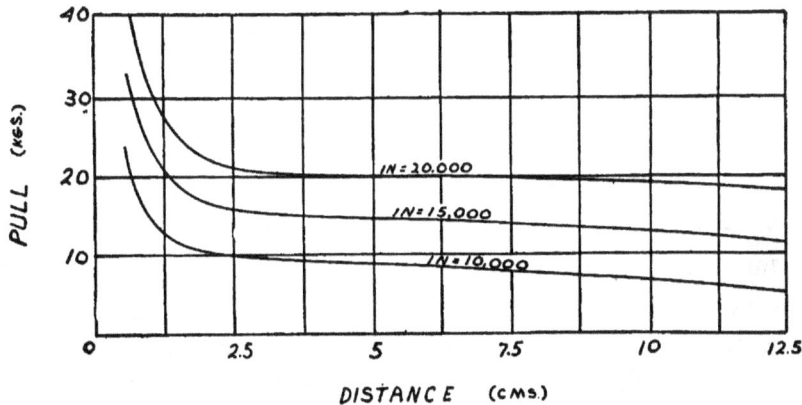

FIG. 79. — Characteristics of Iron-clad Solenoid. $L = 17.8$.

attraction is very great. The solenoid from which Fig. 79 was obtained had no hole through the rear end of the frame.

The general dimensions of the coils and plungers were as follows:

Fig.	L	r_a	A
75	4.6	1.3	1.6
76	8.0	1.8	3.4
77	11.4	2.4	5.1
78	15.2	3.1	9.6
79	17.8	3.5	11.5

L and r_a are in centimeters and A in square centimeters.

From the foregoing, it may be generally stated that the effective range of an iron-clad solenoid is approximately $0.6\,L$; that is, six tenths of the length of the winding. The distances in the charts are measured from the inner attracting face of the iron frame.

47. PRECAUTIONS

The windings of very long solenoids should be divided into sections; the reason for this will be found in Chap. XV, p. 201.

The cross-sectional area of the plunger will depend upon the quickness of action desired. Although the action of a solenoid is naturally sluggish, owing to the fact that the field due to the moving plunger sets up counter-electromotive forces in the winding, a fairly rapid action may be obtained by keeping the cross-sectional area of the plunger small, and making the ampere-turns relatively higher. This method, however, is rather expensive, where the solenoid is to be in circuit long.

The solenoid is unique in that a direct pull may be obtained over a long range of action. Generally speak-

ing, it does not pay to use a lever or analogous mechan-
ism to increase the range, for while the cost of the
winding, frame, and plunger will vary directly with the
length, for a given cross-section of winding, frame, and
plunger, it must be remembered that, for the same
amount of wire, that wound on a small radius will produce more turns than with a larger average radius.
Hence, for the same amount of electrical energy more
work may be obtained, with the same amount of
material, from a long solenoid than may be obtained
from a short one.

In this connection, it might be argued that it would
pay to use a lever in connection with a long solenoid,
to increase the pull, though reducing the range, but the
cost and bother of the lever will seldom compensate for
the advantage gained.

CHAPTER VIII

PLUNGER ELECTROMAGNETS

48. PREDOMINATING PULL

AN iron-clad solenoid provided with a *stop*, as described at the end of Chap. VI, p. 99, is known as a *Plunger Electromagnet.* (See Fig. 80.)

While in the simple and iron-clad solenoids the predominating pull is between the magnetizing force due to the

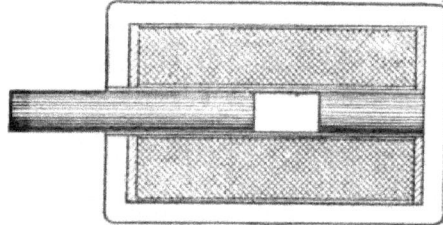

FIG. 80. — Plunger Electromagnet.

current in the winding and the flux in the iron plunger, the pull due to the flux in the plunger and stop predominates in the plunger electromagnet.

49. CHARACTERISTICS

The curves *a* and *b* in Fig. 81 are the same as in Fig. 73 and are due to the iron-clad solenoid in Fig. 72 with the 30.5-cm. coil. Curve *c* was obtained with a stop 25 per cent of the length of the winding, while the stop used in obtaining curve *d* was twice as long; that is, 50 per cent of the winding length. The plunger was 6.45 sq. cm. in cross-section.

These characteristics are particularly interesting as they are obtained from an actual test * made by the

* *Electrical World and Engineer*, Vol. XLV, 1905, pp. 934–935.

author. The magnet had a massive wrought-iron return circuit. The curves e and f are calculated by formula (92).

$$P_m = \left(\frac{IN}{3951\,l}\right)^2 A.$$

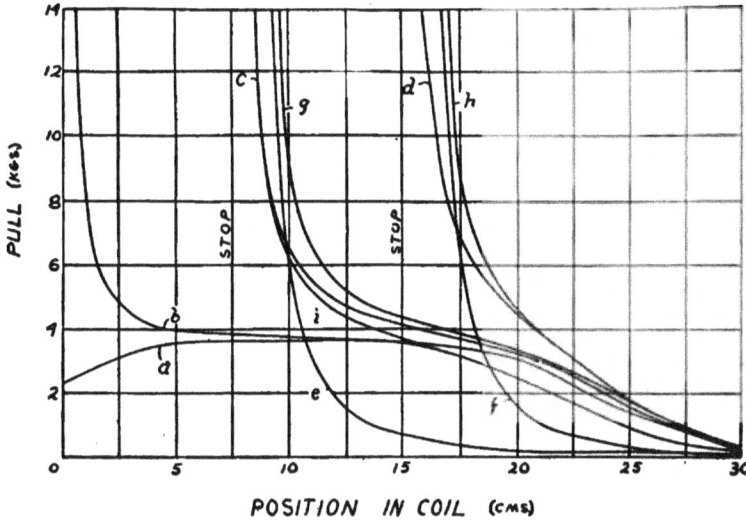

POSITION IN COIL (CMS)

FIG. 81. — Characteristics of Plunger Electromagnet.

Inspection of these curves will show that if the heights of curve e be added to those of curve b, curve g will be the result. Likewise, the addition of the heights of curves f and b will produce curve h.

50. CALCULATION OF PULL

Now, curves g and h are calculated by formula (95).

$$P_s = A\left[\frac{\theta_t(IN - A\lambda)}{981\,L} + \left(\frac{IN}{3951\,l}\right)^2\right].$$

The reason why the actual and calculated values do not coincide is on account of the magnetic reluctance of the iron plunger at so high a density. All the curves in Fig. 81 are due to 10,000 ampere-turns.

By assuming the reluctance to be equivalent to one centimeter length of air-gap, under these conditions, curves g and h will exactly coincide with curves c and d, respectively.

Part of the quantity assigned to reluctance is due to leakage, but it is easier to consider it all as reluctance, if provisions are made for it.   (See p. 39.)

51. Effect of Iron Frame

Excepting for a very short range of action, the reluctance of the iron frame appears to have but little effect at this high density in the plunger and stop.

The curve marked i in Fig. 81 is due to using the same coil, plunger, and stop, as in the test which gave curve c, but with no iron return circuit. A large block of iron was, however, placed at the rear end of the coil.

It will be noticed that the lower part of the curve i tends to follow the lower part of curve a, which would seem perfectly natural, as there was no iron at the mouth of the winding other than the plunger itself, when curves a and i were made.

This would indicate that it is not necessary to use a very heavy iron frame, and that the magnetic connection between the plunger and the frame at the mouth of the winding, is not a matter of much importance for a high m. m. f.

In the foregoing, the flux density was very high for short air-gaps. It is evident, however, that for a short range of action, it is more important to work with low-flux densities.   Hence, for short air-gaps and with about 75 per cent saturation, formula (95) may be reduced to

$$P_s = A\left[\frac{IN}{490\,L} + \left(\frac{IN}{3951\,l}\right)^2\right]. \qquad (96)$$

In this, the value of θ_t is made maximum, $i.e.$ 2, and λ reduced to zero. As the reluctance of the air-gap will be practically all the reluctance in the circuit, only a slight allowance may be made for leakage.

While no exact rule for the reluctance (including leakage, and the bulging of the lines around the air-gap) may be set forth, unless an exact knowledge of the iron characteristics are known, the statement regarding allowances for solenoids, given in Art. 37, p. 80, will hold for plunger electromagnets also.

52. Most Economical Conditions

The proper flux density may be determined by a method due to Mr. E. R. Carichoff. To quote from one of his articles* " The main facts that seemed to the writer as useful are that there is a certain length of air-gap for any given magnet of uniform cross-section where the pull between armature and poles is lessened if the polar area is either increased or diminished, and that the pull under these conditions, multiplied by the length of the air-gap, is greater than the pull with any other air-gap multiplied by the length of the latter."

The following explanation of his method † is herewith reproduced :

* *The Electrical World*, Vol. XXIII, 1894, pp. 113-114.
† *The Electrical World*, Vol. XXIII, 1894, pp. 212-214.

Let us assume, for example, that the curve in Fig. 82, *OECD*, represents the iron, and *OF* the air-gap

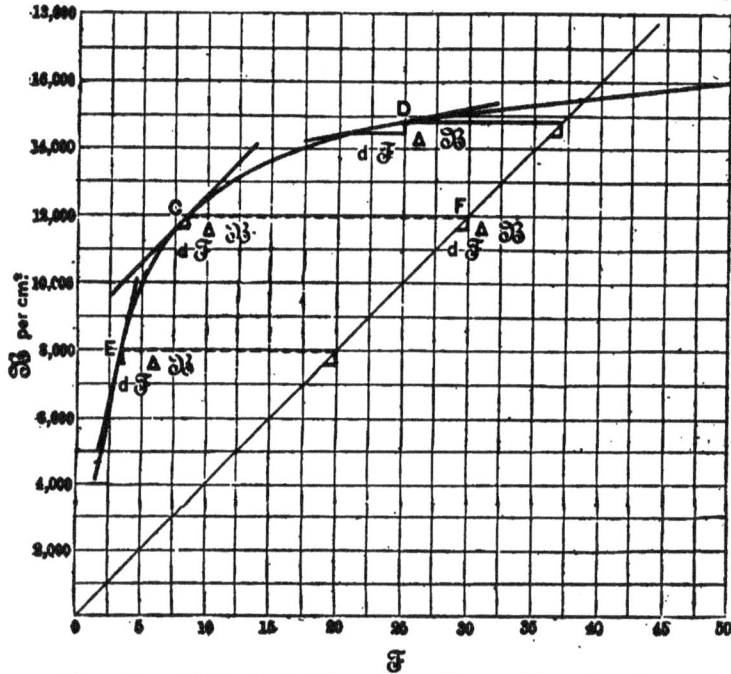

FIG. 82.—Method of determining Proper Flux Density.

characteristic, and that we are working at the point *C* on said curve. Reduce the polar area by dA, and suppose that the force is reduced for an instant by dF, so that the induction in the air-gap is still $\mathscr{B}$ and that in the iron is reduced by $\Delta\mathscr{B}$. Since a tangent drawn to the curve at *C* is parallel to the line *OF*, we see that the force necessary to produce a change $\Delta\mathscr{B}$ in the induction is the same in both iron and air-gap. Therefore, if $d\mathscr{F}$ produces, in iron, a change $\Delta\mathscr{B}$, it can produce $\frac{1}{2}\Delta\mathscr{B}$, say $d\mathscr{B}$, in both.

It is evident that with the above assumptions

$$\frac{\Delta\mathscr{B}}{\mathscr{B}} = \frac{dA}{A} \text{ and } d\mathscr{B} = \frac{1}{2}\mathscr{B}\frac{dA}{A}.$$

With area A and induction $\mathcal{B}$ the pull is proportional to $\mathcal{B}^2A$.

With the area $(A - dA)$ and the induction $\mathcal{B} + d\mathcal{B}$, we have $(\mathcal{B} + d\mathcal{B})^2(A - dA)$ proportional to pull.

Then $\qquad \mathcal{B}^2A < = > (\mathcal{B} + d\mathcal{B})^2(A - dA),$ $\qquad$ (97)

if $\qquad\qquad \dfrac{1}{2}\mathcal{B}\dfrac{dA}{A} < = > d\mathcal{B}.$

Since for this case $d\mathcal{B} = \dfrac{1}{2}\mathcal{B}\dfrac{dA}{A}$, the two sides of the equation are the same and the pull is the same when the polar area is A, and when it is $A - dA$. In the same way it can be shown that the pull is the same where the polar area is A and $A + dA$.

From this we draw the conclusion that when the air-gap reluctance is expressed by a line parallel to the tangent drawn at the part of the iron characteristic where we are working, there is no gain by either increasing or decreasing the polar area. This is evidently the condition of maximum efficiency, as we shall see from further considerations.

Keep the same characteristics and suppose our ampere-turns put us at the point D on the iron curve. Decrease the polar area by dA, and for an instant the force by $d\mathcal{F}$, so that $\mathcal{B}$ is still the induction in the air-gap. $d\mathcal{F}$ is now much greater than before necessary to make the change $\Delta\mathcal{B}$, so that it will produce a change in the whole circuit by an amount $d\mathcal{B}$ referred to air-gap, where $d\mathcal{B}$ is greater than $\frac{1}{2}\Delta\mathcal{B}$, or greater than $\dfrac{1}{2}\mathcal{B}\dfrac{dA}{A}$. This condition makes the right-hand member of equation (97) greater than the left-hand member, showing that the pull is increased by decreasing the polar area. Increase the polar area by dA, keeping other conditions as above, and equation (97) shows that the pull is decreased. Therefore, if our iron curve shows that $\dfrac{\Delta \mathcal{EB}}{d\mathcal{F}}$ is greater than the same function for the air-gap, the air-gap reluctance should be increased until the two functions are the same.

Again, at the point E we find that $d\mathcal{F}$ is less than in the case first cited, and if the same reasoning is followed, it is found that the pull is increased by increasing the polar area.

Instead of increasing or decreasing the polar area,

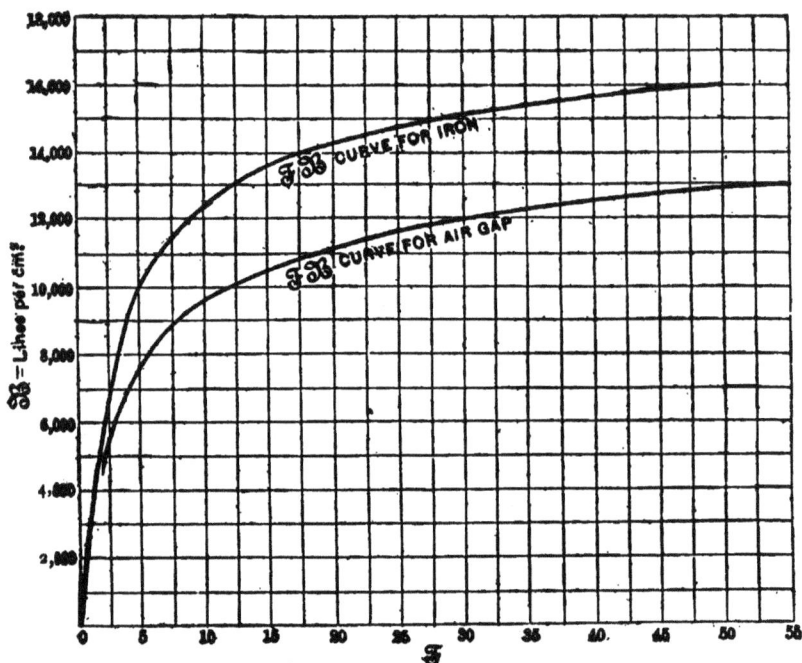

Fig. 83. — $\mathcal{FB}$ Curves for Iron and Air-gap.

the same results are obtained by shortening or lengthening the air-gap. When this is done, so that the function $\dfrac{\Delta\mathcal{B}}{d\mathcal{F}}$ is the same for iron and air-gap, the pull multiplied by the distance is a maximum. This being the case, the best results are obtained by using the proper air-gap; and if this distance is not the travel required, a lever can be used to multiply or divide with.

With these principles in view, it is easy to get the most efficient possible arrangement. It is also easy to predetermine a magnet to pull required amounts in more than one point of its travel.

By drawing tangents to all points of the iron curve and lines from the origin parallel to those tangents, and intersecting these by horizontal lines, we get the curve of proper air-gap for all points of iron curve cut by said horizontal lines. (See Fig. 83.)

To prove that the air-gap thus determined gives the maximum amount of work, where work equals initial pull times length of air-gap, change l by dl; then

$$\frac{\Delta\mathcal{B}}{\mathcal{B}} = \frac{dl}{l}.$$

We see that for the point F

$$\frac{d\mathcal{F}}{\Delta\mathcal{B}} = \frac{d\mathcal{F}}{\Delta\mathcal{B}}$$

and $$d\mathcal{B} = \tfrac{1}{2}\,\Delta\mathcal{B},$$

and for points to the right of F,

$$d\mathcal{B} > \tfrac{1}{2}\,\Delta\mathcal{B},$$

and for points to the left of F,

$$d\mathcal{B} < \tfrac{1}{2}\,\Delta\mathcal{B}.$$

In the three cases the work will be proportional respectively to

$$\mathcal{B}^2 l. \tag{1}$$

$$(\mathcal{B} - d\mathcal{B})^2(l + dl). \tag{2}$$

$$(\mathcal{B} + d\mathcal{B})^2(l - dl). \tag{3}$$

$$(1) - (2) = +, \text{ for } d\mathcal{B} > \frac{1}{2}\mathcal{B}\frac{dl}{l}.$$

$$(1) - (3) = +, \text{ for } d\mathcal{B} < \frac{1}{2}\mathcal{B}\frac{dl}{dl}.$$

In Fig. 84 is shown the result of a test made by the author with the $L = 17.8$, $r_a = 3.5$, $A = 11.5$, solenoid, sur-

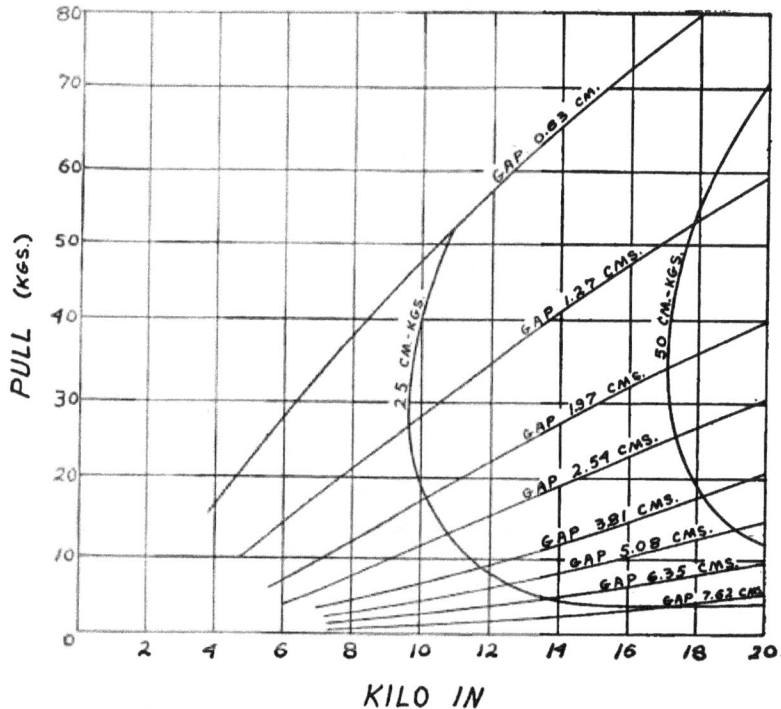

Fig. 84. — Air-gaps for Maximum Efficiency.

rounded by a cast-iron frame. The stop extended throughout nearly one half the length of the winding, which is the position of maximum pull, as will be shown presently.

Reference to this diagram will show that the most economical condition, with this particular plunger electromagnet, is met with a 1.27-cm. air-gap, for 25 cm.-kg. of work, while for 50 cm.-kg. the length of the air-gap would be slightly greater. This shows plainly that only the best grades of iron or steel should be used where high efficiency is desired.

53. Position of Maximum Pull

Figure 85 is the result of a test of the above magnet, with a constant air-gap of 1.27 cm. and various lengths of stops. The position of gap is measured from the inner end of the frame.

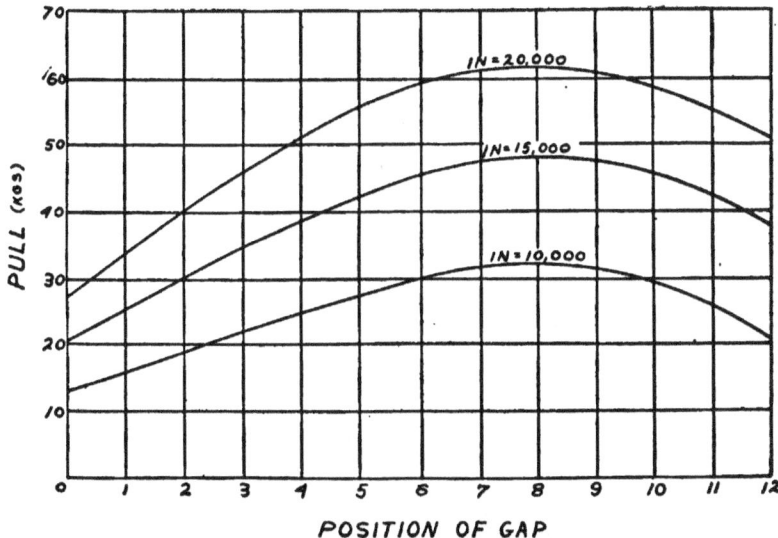

FIG. 85. — Test showing Position of Air-gap for Maximum Pull.

It is self-evident that the maximum pull is obtained when the center of the air-gap is near the center of the winding, assuming the winding to be equally distributed between the limits of the iron frame.

54. CONED PLUNGERS

It has been shown that there is a certain relation between polar area and length of air-gap which will produce the most economical conditions ; hence, if the length of the air-gap is to be increased, the polar area must be increased also. However, if the diameter of the plunger be increased, there may not be enough room for the winding if the space is limited.

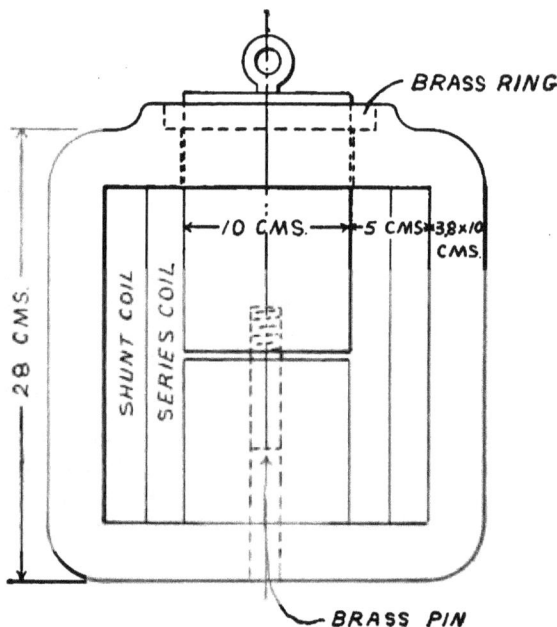

FIG. 86. — Flat-faced Plunger and Stop.

From another article by Mr. E. R. Carichoff,* part of the following is taken:

Figure 86 shows the dimensions of a cast-steel electromagnet, which, with an air-gap of 0.316 cm., pulls 545 kg. when the exciting power is 4800 ampere-turns in

* *The Electrical World*, Vol. XXIV, 1894, p. 122.

either shunt or series coil. This gap is, for this particular case, the one which, multiplied by the pull of 545 kg., gives a maximum, or 173 cm.-kg. In case a movement over 1.27 cm. instead of 0.316 cm. is required, the plunger of the magnet may be attached to the short end of a 4 to 1 lever, at the long end of which the travel will be 1.27 cm., and the pull 136 kg., or, as before, 173 cm.-kg. If, however, the plungers are simply separated 1.27 cm., the initial pull becomes about 60 kg., which, multiplied by 1.27 cm., gives 76 cm.-kg., or less than one half of what can be obtained by using the lever.

In this it is assumed that the force to overcome is fairly constant, or but slightly increasing. Of course, if the force to be overcome varies directly with the pull of the magnet, there is no need to bother about the air-gap.

It is possible to get a direct pull of 136 kg. at a distance of 1.27 cm. by using the form of magnet in Fig. 87, which was suggested and adapted by Lieutenant F. J. Sprague. The only change is in the air-gap. With the angles there shown for the male and female

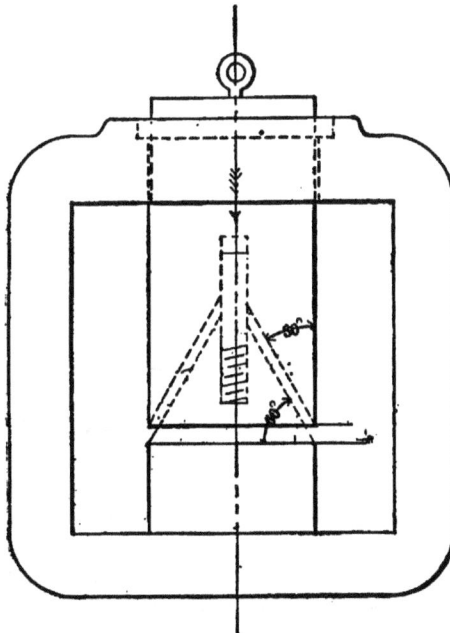

Fig. 87. — Coned Plunger and Stop.

cone plungers it is readily seen that the area of air-gap

is approximately doubled, and its length also doubled, while the reluctance is approximately the same as that of the arrangement in Fig. 86.

It is further readily seen that the travel of plunger is 1.27 cm. instead of 0.316 cm., as in Fig. 86.

As the reluctance of the circuit is approximately the same as in Fig. 86, the total number of magnetic lines is approximately the same, but as these lines are distributed over an air-gap of double the area, and the force makes an angle of 60 degrees with the direction of travel, the pull is only 136 kg. at 1.27 cm. This pull of 136 kg. times 1.27 cm. is, however, 173 cm.-kg., as before.

To change polar area and travel, keeping the reluctance approximately constant: Assume the approaching poles to be two cylinders with male and female conical extremities, as in Fig. 88.

If A is the cross-section, and d the travel, and l the proper length of air-gap for a right section, A' is the polar area, l' is the air-gap length, and d' the travel when the angle at the base of the cone becomes α.

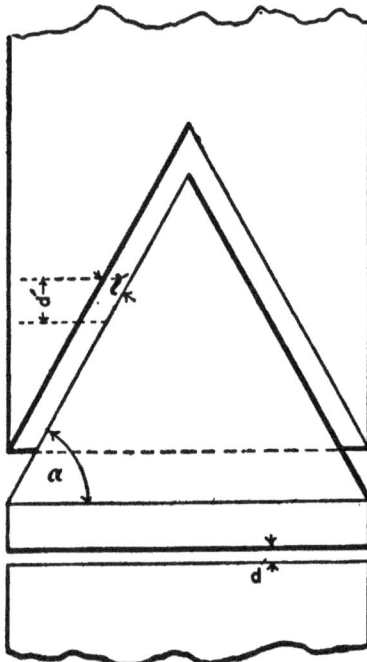

FIG. 88. — Comparison of Dimensions and Travel of Flat-faced and Coned Plungers and Stops.

Condition of constant reluctance is $\dfrac{l'}{A'} = \dfrac{d}{A}$.

But $\qquad A = A' \cos \alpha$ and $l' = d' \cos \alpha.$

Therefore, $\qquad \dfrac{d' \cos \alpha}{A} = \dfrac{d}{A' \cos \alpha},$

whence $\qquad \cos^2 \alpha = \dfrac{d}{d'}$ and $\dfrac{A^2}{A'^2} = \dfrac{d}{d'}.$

Thus, if the polar area is doubled in this way, the travel is increased four times.

It is also seen that l' differs more or less from the true length of air-gap according to position of conical surfaces.

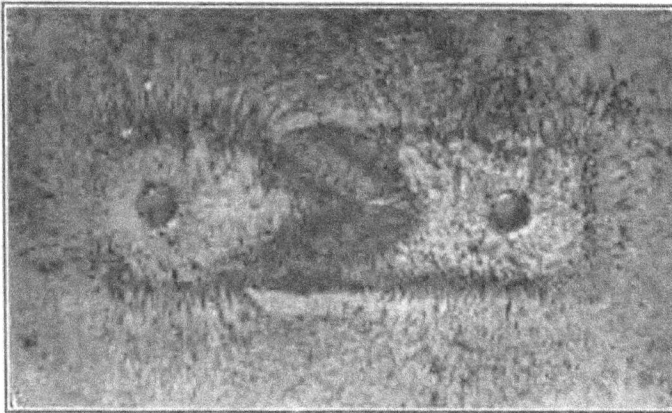

FIG. 89. — Flux Paths between Coned Plunger and Stop.

Figure 89 shows the flux paths for a V-shaped air-gap between two pieces of iron, which may be considered as a good representation of the conditions in the air-gap of the plunger electromagnet in the article above quoted.

The effect of changing the angles of pointed plungers is shown in Fig. 90, which curves are due to the $L = 17.8$, $r_a = 3.5$, $A = 11.5$, plunger electromagnet previously referred to. The stop extended 35 per cent through

the coil.   These effects are due to 20,000 ampere-turns.
In practice, a larger plunger and lower m. m. f. should
be employed.

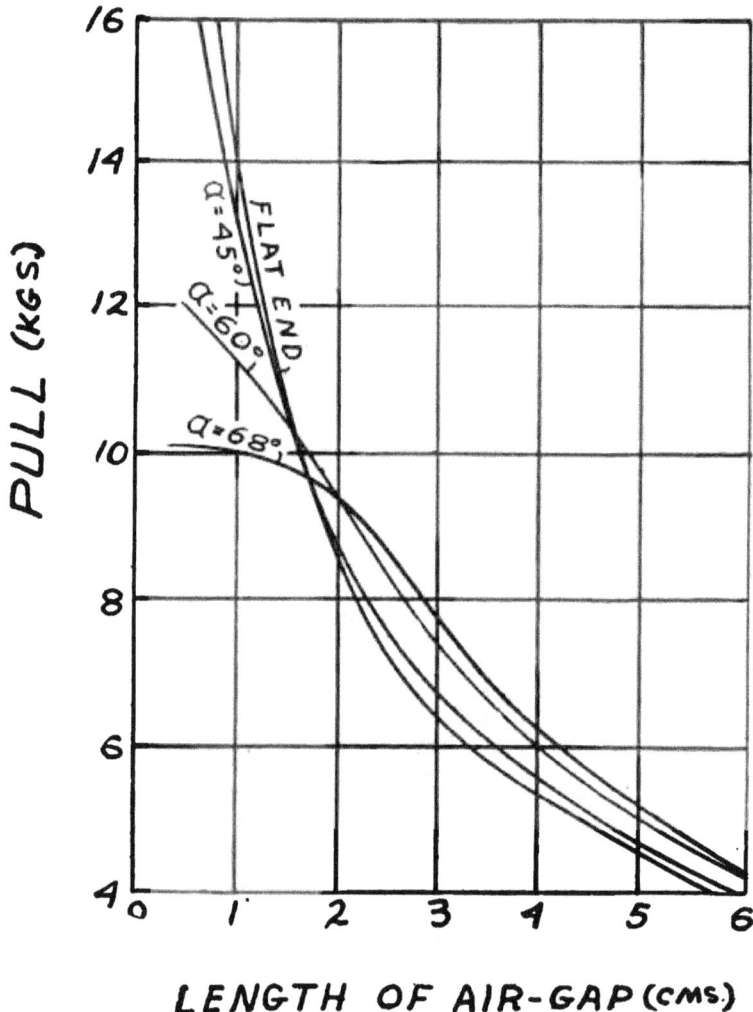

LENGTH OF AIR-GAP (CMS.)

Fig. 90. — Effect of changing Angles.

Mr. W. E. Goldsborough * has described the electromagnet shown in Fig. 91, which is, no doubt, the best

* *Electrical World and Engineer*, Vol. XXXVI, 1900.

application of the coned plunger, as the conical portion of the plunger is the same length as the winding. In

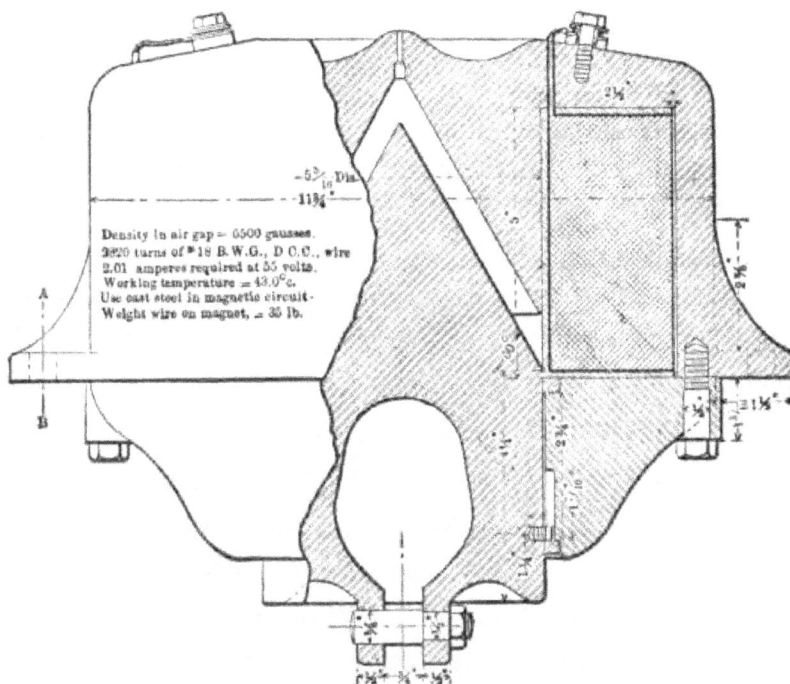

FIG. 91.—Design of a Tractive Electromagnet to perform 400 cm.-kg. of Work.

this case the range is so short, and the attracting area so great, that the solenoid effect may be entirely neglected, and formula (68) may be employed. (See p. 76.)

55. TEST OF A VALVE MAGNET

The following test,* described by Mr. C. P. Nachod, is of interest.

This magnet, shown in Fig. 92, has a short stroke, and is designed to operate an air-valve by means of a

* *Electrical World and Engineer*, Vol. XLVI, 1905, p. 1071.

brass rod (not shown) passing through the hole in the core. The core and armature are of Norway iron; the core has an outside diameter of 3.17 cm. and has a 0.95-cm. hole through it. The magnet is of the iron-clad type with cast-iron shell and top. There is a cylindrical clearance gap between the armature and top, which, being invariable, has not been considered.

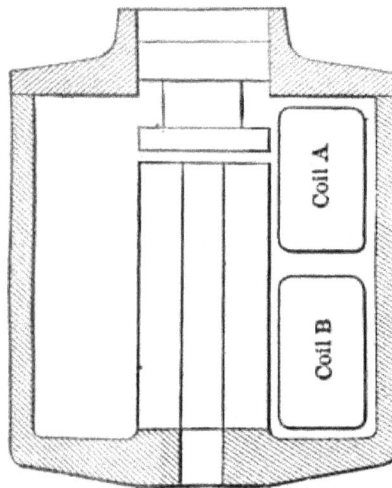

FIG. 92. — Valve Magnet.

There are two exciting coils, *A* and *B*. The former, having 1860 turns of No. 25 wire, is used as an operating coil to draw down the armature; and the latter, of 1765 turns of No. 26 wire, serves as a retaining coil to hold it down against the air-pressure, after the circuit in the other coil is broken. Coil *A*, as shown in Fig. 92, is located so as to encircle the working air-gap, which is not the case with *B*.

The test was made with a simple traction permeameter, definite air-gaps being obtained by measured brass washers.

Figure 93 shows the pull produced with varying excitation for each coil separately, with a constant air-gap of 0.41 mm. In the curve for coil *A* the range of magnetic densities is sufficient to show clearly the change in permeability of the iron circuit. For so small an air-gap the difference between the pulls produced by the two coils with the same magnetizing force is remarkably great. The ratio of these pulls has been plotted, and the curve shows that coil *B* is

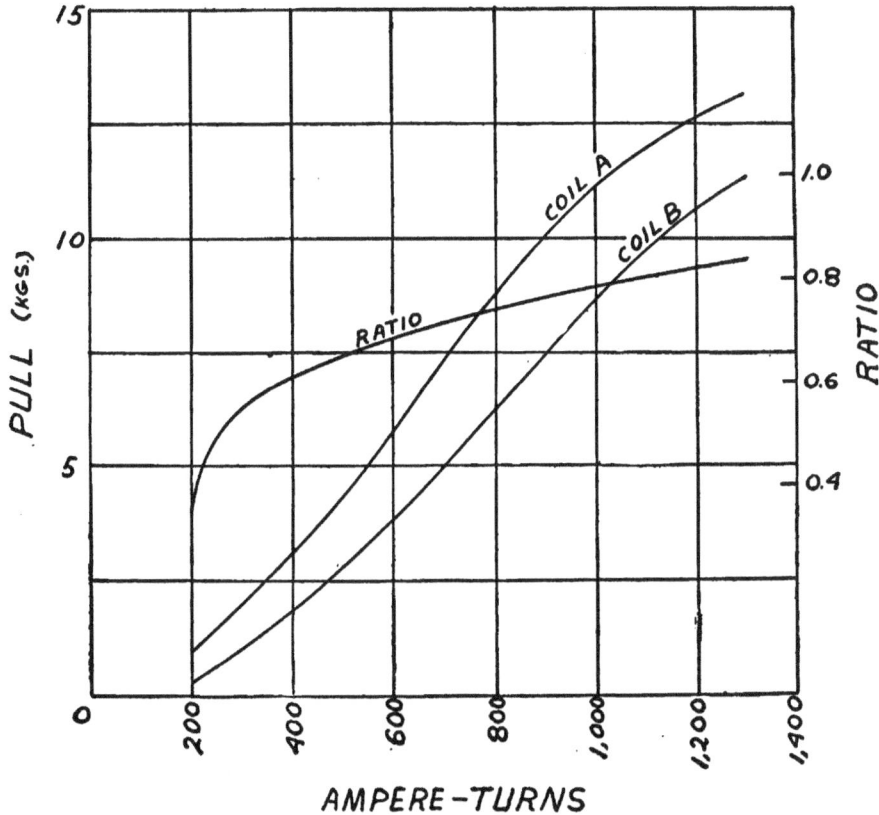

FIG. 93. — Characteristics of Valve Magnet.

only from 0.4 to 0.8 as effective as *A*, due solely to its position on the core, which permits a larger magnetic leakage.

Figure 94 shows the same relations with an air-gap of 2.05 mm. or five times as great as in the previous case. With this gap the range of flux density is not great enough to produce an inflection point in either curve; and the ratio of pulls for the same excitation is more nearly constant, averaging a little over 0.7, but decreasing with decreasing density.

From the curves it appears that if it were desired to connect the coils in series so that the magnetic effect

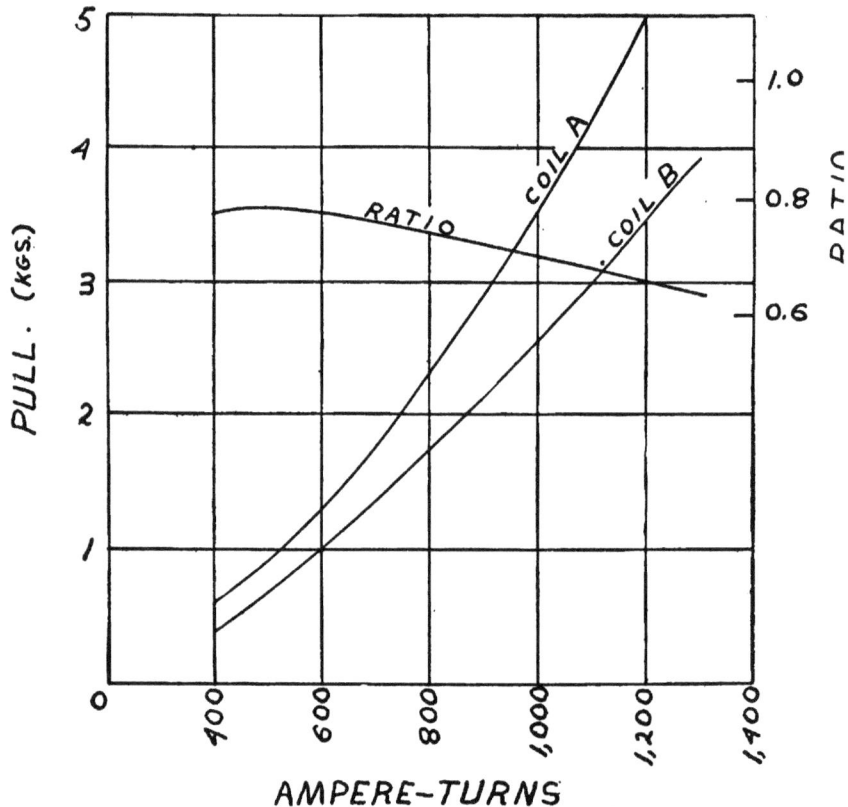

FIG. 94. — Characteristics of Valve Magnet.

of B would neutralize that of A, it would be possible to do so for only one, instead of all current values, and that such a neutralizing coil for all current values should have the same axial length and slide within or without the other.

56. COMMON TYPES OF PLUNGER ELECTROMAGNETS

The general design of the frames of plunger electro-magnets is optional, providing the flux passes from the frame to the plunger uniformly on all sides, so as not to attract the latter sidewise.

The plunger usually travels within a brass tube placed within the insulated winding.

FIG. 95. — Horizontal Type Plunger Electro-magnet.

Any form of guide consisting of

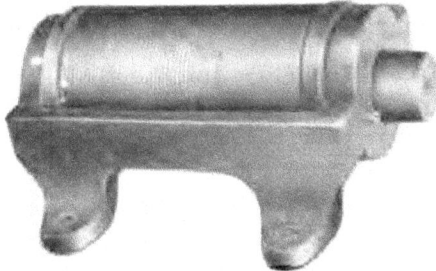

FIG. 96. — Horizontal Type Plunger Electromagnet.

non-magnetic material may be employed.

In Figs. 95 to 97 there are shown several common types. The magnet in Fig. 98 has

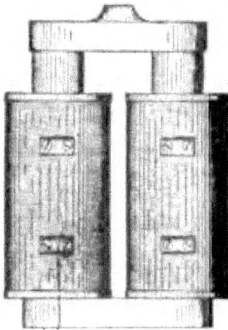

FIG. 97. — Vertical Type Plunger Electromagnet.

FIG. 98. — Two-coil Plunger Electromag-net.

no outer frame; nevertheless, it has a closed magnetic circuit outside of the working air-gap.

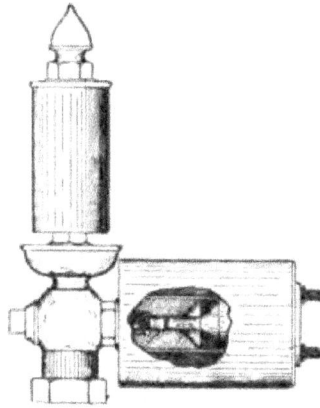

FIG. 99. — Pushing Plunger Electromagnet.

57. PUSHING PLUNGER ELECTROMAGNET

Where a pushing effect is desired, the plunger electromagnet may be inverted, and a brass rod fastened to the plunger and passed through the stop, as in Fig. 86, Art. 54.

A magnet of this type is shown in Fig. 99 attached to a whistle valve.

58. COLLAR ON PLUNGER

By placing a collar on the plunger outside of the frame, as in Fig. 100, the pull may be considerably increased.

FIG. 100. — Electromagnet with Collar on Plunger.

This is due to the flux which passes from the frame to the plunger, at the opening through the frame.

CHAPTER IX

ELECTROMAGNETS WITH EXTERNAL ARMATURES

59. Effect of placing Armature Outside of Winding

In Fig. 101* is shown the effect of ascertaining the pull due to the magnet in Fig. 102 (also referred to in

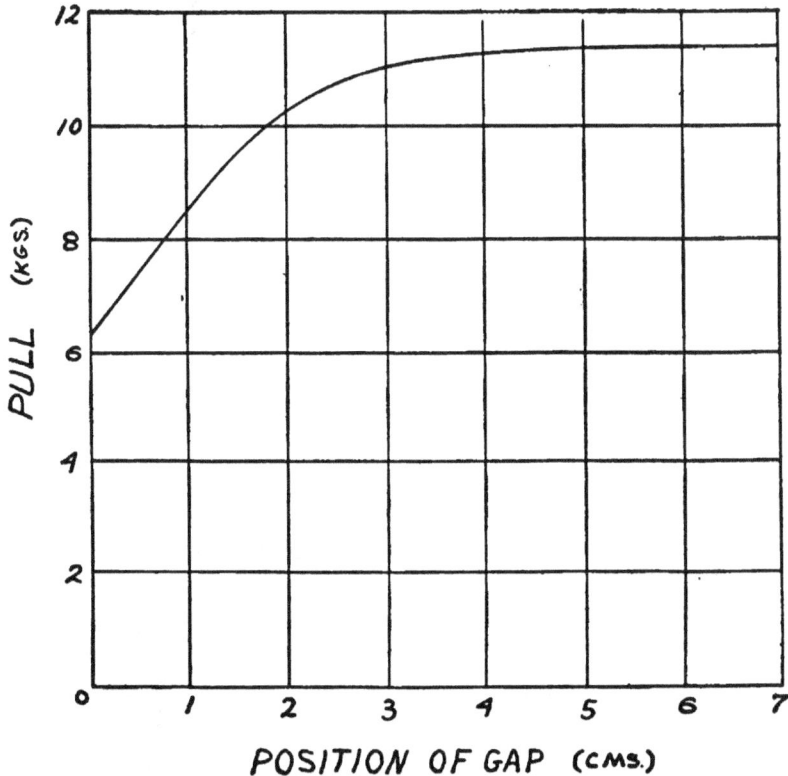

FIG. 101. — Characteristics of Test Magnet.

*American Electrician, Vol. XVII, 1905, pp. 299–302.

Art. 42), by using two cores of the same length, separated by a brass disk 0.89 mm. thick, which represented the air-gap (the permeability of brass being the same as for air). The ampere-turns were 6300.

It will be observed that when the abutting ends of the cores were at the end of the winding, the pull was only approximately one half of what it was at the center of the coil.

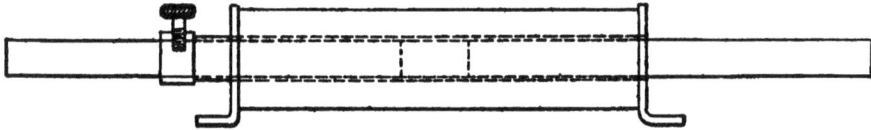

FIG. 102. — Test Magnet.

This effect is easily explained. When the abutting ends are at the mouth of the winding, the leakage is so great that only about seven tenths of the total flux passes through the working air-gap.

60. BAR ELECTROMAGNET

When the core and coil are in the relative positions as shown in Fig. 103, they constitute the *Bar Electromagnet*. Its field is very similar to that of the bar permanent magnet.

FIG. 103. — Bar Electromagnet.

To obtain good results from this type the ends of the core should be bent as in Fig. 104; but if these *limbs* be very long, the leakage between them will be great. A similar effect may be obtained by bending the armature instead of the core.

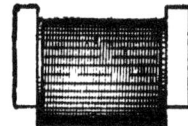

FIG. 104. — Electromagnet with Winding on Yoke.

61. Ring Electromagnet

The effect of bending a bar electromagnet into a circle is shown in Fig. 23, p. 37.   This is called a *Ring Electromagnet*.   This type has the minimum leakage, but if the radius of the ring be small, the turns constituting the exciting coil tend to crowd at the inside of the ring, which increases the total length of wire and reduces the ampere-turns for a given voltage; while if the radius be large, the reluctance of the magnetic circuit will be correspondingly great.

62. Horseshoe Electromagnet

The natural method, then, is to bend the core in the form of U and place a coil on each limb as in Fig. 105. This form is, however, rather inconvenient to make, and the bend takes up too much room, as a rule; therefore, the practical horseshoe elec-

Fig. 105. — Horseshoe Electromagnet

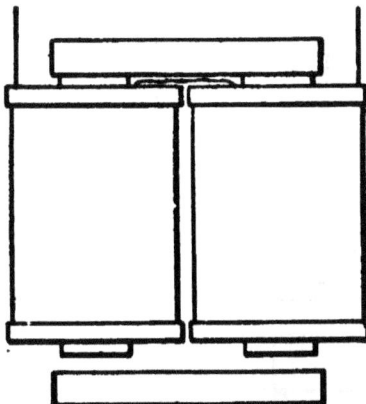

Fig. 106. — Practical Horseshoe Electromagnet.

tromagnet is made of three pieces, besides the armature. In this form, Fig. 106, the wire is wound directly on to the insulated cores, and the latter are then fastened to the yoke, or "back iron," as it is often called.

In Fig. 107 is shown a modification of the horseshoe electromagnet. This is also comparable with the bar electromagnet with one of its core ends bent around near the opposite end of the core.

FIG. 107. — Modified Form of Horseshoe Electromagnet.

This is not so economical as the two-coil electromagnet, as there cannot be so many ampere-turns for the same amount of copper for one coil of large diameter as with two coils of smaller diameter, for the same total current, as is pointed out in Art. 164, p. 284.

63. TEST OF HORSESHOE ELECTROMAGNET

The result of a test of the magnet in Fig. 108 is shown in Figs. 109 and 110.

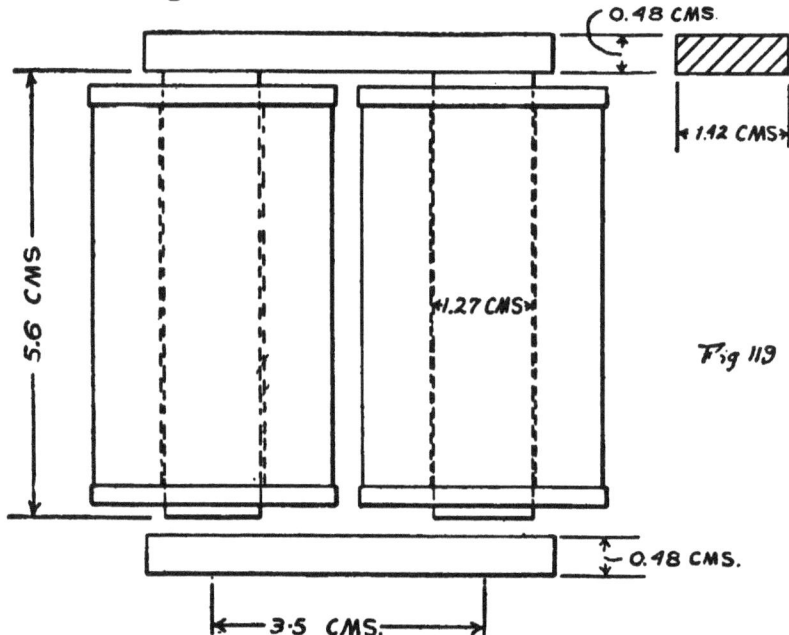

FIG. 108. — Experimental Electromagnet.

It will be observed that the pull is strong for very short air-gaps, but weak when the armature is removed a short distance from the cores.

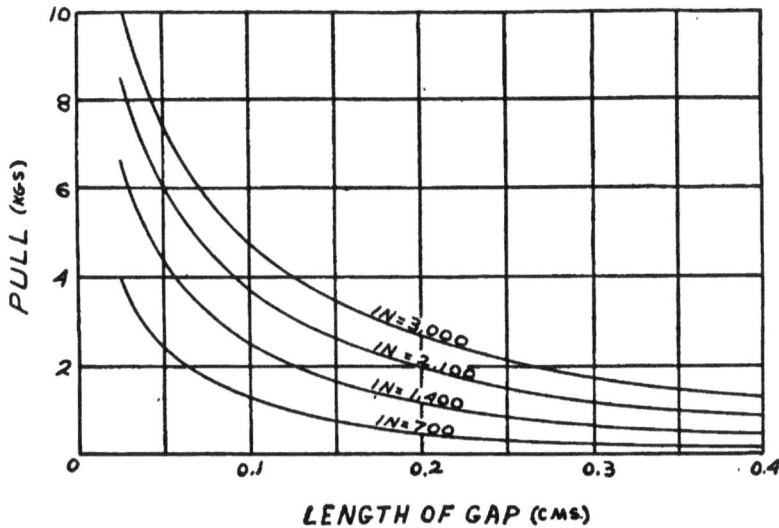

FIG. 109. — Characteristics of Horseshoe Electromagnet.

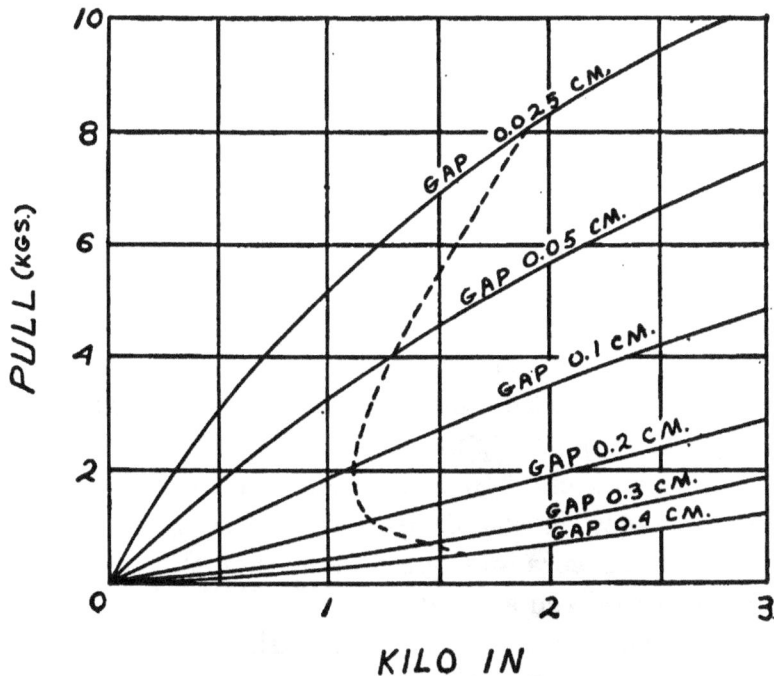

FIG. 110. — Relation of Work to Length of Air-gap.

Since, in this case, there are two air-gaps, the total air-gap is twice that indicated in Figs. 109 and 110.

The dotted curve in Fig. 110 represents 0.2 cm.-kg. of work. It will be observed that the most economical condition for this amount of work is met with 0.1-cm. (1 mm.) air-gap.

64. Iron-clad Electromagnet

This type may be divided into two classes, one of which is usually of small dimensions and consists of a piece of

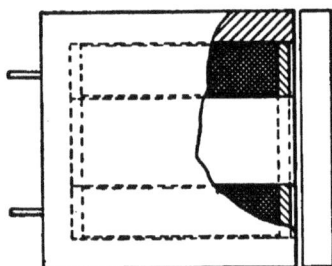

FIG. 111. — Iron-clad Electromagnet.

soft-iron tubing with a soft-iron disk at one end to which the core of the spool containing the winding is fastened, Fig. 111, or else it is made from a solid piece of iron, or steel casting, by turning a groove to receive the exciting coil. The latter construction is usually adhered to in large magnets.

Small iron-clad electromagnets are used extensively in telephone switchboard apparatus where the range of action is not great and the duration of excitation is brief. The fact that it is not materially affected by external magnetic influence makes it particularly adapted for this purpose.

Its use is limited to cases where the range or attracting distance is small, owing to the great leakage between the core and outer shell when the armature is removed for even a small distance.

The iron-clad electromagnet is also employed where a very strong attraction is desired when the armature is in actual contact with the polar surfaces. Since in this case the air-gap is exceedingly small, so many ampere-turns are not required, and the magnetic circuit may be made very short.

Electromagnets of this class are fitted with tool-holding devices and may be rigidly held to the beds or frames of machines in any desired position by simply turning on the current. They are also employed for gripping iron pipe, pig, etc., while the latter are being hoisted.

65. LIFTING MAGNETS

Electromagnets of the iron-clad type have, in recent years, come into general use in large industrial plants for handling iron and steel of every description, both hot and cold.

Some of the principal types are shown in the following illustrations.*

Figure 112 shows a " skull-cracker " at work reducing a scrap heap. The large steel ball is gripped by the magnet, both lifted by the crane, the current turned off, and the ball drops, reducing the large pieces of scrap to small pieces which may be easily handled. The same magnet is employed to place the large pieces in position and to remove the broken pieces. These latter operations are similar to that shown in Fig. 113, wherein scrap is being unloaded from cars by means of the magnet.

Another type, known as *Plate and Billet Magnet*, is shown in Fig. 114.

In Fig. 115 is shown an *Ingot Magnet* lifting an ingot mold. There are many other uses for these magnets.

* Electric Controller and Supply Co.

Fig. 112. — Skull Cracker.

Fig. 113. — Lifting Magnet.

Fig. 114. — Plate and Billet Magnet.

FIG. 115. — Ingot Magnet.

The cost of handling the melting stock used by open-hearth furnaces from cars to stock pile, or from stock pile to charging boxes, has been reduced from approximately eight cents a ton by hand methods to two cents a ton by the use of magnets in connection with suitable cranes.

66. CALCULATION OF ATTRACTION

The relation between ampere-turns and pull, for the air-gap alone, is

$$P = A_g \left(\frac{IN}{3951 \, l_g} \right)^2, \qquad (98)$$

wherein P is the pull in kilograms, A_g the cross-sectional area of each gap (if more than one), and l_g the length of the air-gap in centimeters.

Transposing,

$$IN = 3951 \, l_g \sqrt{\frac{P}{A_g}}. \qquad (99)$$

For electromagnets of the type shown in Fig. 107 the complete formula (including leakage, see p. 39) is

$$IN = 3951 \, l_g \sqrt{\frac{P}{A_g}} \left(\frac{l_c l_g}{2 \, \mathcal{R}_l A_g} + 1 \right), \qquad (100)$$

while for the horseshoe type

$$IN = 3951 \, l_g \sqrt{\frac{P}{A_g}} \left(\frac{l_c l_g}{\mathcal{R}_l A_g} + 1 \right), \qquad (101)$$

wherein l_c = length of each core or limb in centimeters and $\mathcal{R}_l$ is the reluctance between parallel cores, limbs, or surfaces per centimeter length.

These formulæ are, however, only approximately correct.

Where the end area of the core is not the same as the attracting surface of the armature, A_g will be the average between the two areas.

An electromagnet designed to attract its armature through a given distance may be made more effective by increasing the attracting areas of either the core or armature or both. This may preferably be accomplished, as in Fig. 116.

The lifting power of electromagnets of the class described in Art. 65 may only be determined by experience, since it is impossible to predetermine the attracting areas and air-gaps of the scrap to be lifted.

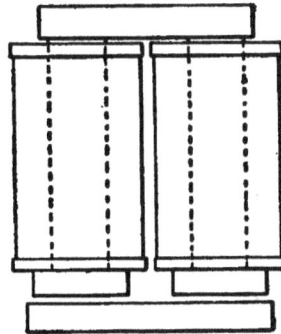

FIG. 116.—Method of Increasing Attracting Area.

There is a peculiar phenomenon in connection with these magnets. Before the material, which may be considered as the armature, comes in actual contact with the polar surfaces, the greater the relative surfaces, the greater will be the attraction; but, after actual contact, the attraction is greater when the polar or attracting surfaces are smaller.

Consider the magnet in Fig. 117. When the armature is in actual contact with the cores, the attraction is greater

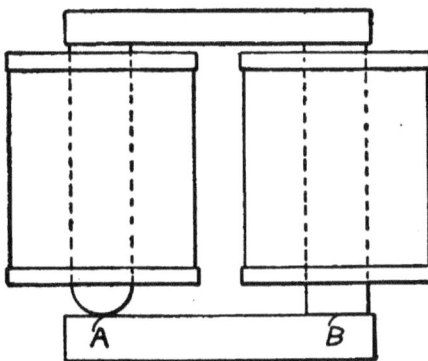

FIG. 117.—Electromagnet with Flat-faced and Rounded Core Ends.

at A than at B. This may be explained as follows: The attraction is proportional to $\mathcal{B}^2 A$.

Now, $$\mathscr{B} = \frac{\phi}{A}.$$ (41)

Hence, $$\mathscr{B}^2 A = \frac{\phi^2}{A^2} A = \frac{\phi^2}{A}.$$

If it is assumed that the flux ϕ is constant, it will be evident that the smaller the value of A the greater will be the attraction.

67. POLARITY OF ELECTROMAGNETS

In practice the exciting coils of horseshoe electromagnets are all wound in the same direction. After the spools are mounted on the yoke, or "back iron," the inside terminals of the coils are connected together, leaving the outer terminals for making connection to other apparatus. Sometimes the coils are connected in multiple. In such cases the inside terminal of each coil must be connected to the outside terminal of the other.

A little reflection will show why the above methods are necessary in order that both coils may have the proper polarity. The relative directions of current and flux are shown on p.24, and this rule may be applied to coils by the following analogue. Consider the direction in which an ordinary screw is turned to be the direction of the current, and the direction of travel of the screw to be the direction of flux.

68. POLARIZED ELECTROMAGNETS

A *Polarized Electromagnet* is a combination of a permanent magnet and an electromagnet. Normally the whole magnetic circuit is under the influence of the permanent magnet alone. When a current flows through the coils of the electromagnet, the polarity of the cores due to the permanent magnet may be augmented; partly or wholly neutralized, or even reversed. Figures 118 and 119 show how the polarization is usually effected. In both cases the armatures and cores are of soft iron. In Fig. 118 the armature is pivoted at the center, and the cores are connected to a soft-iron yoke, the whole being influenced by the permanent magnet, as shown. This type is extensively used in telephone ringers. The type illustrated in Fig. 119 is that commonly used in telegraph apparatus.

FIG. 118. — Polarized Striker Electromagnet.

In this case the armature is pivoted at one end.

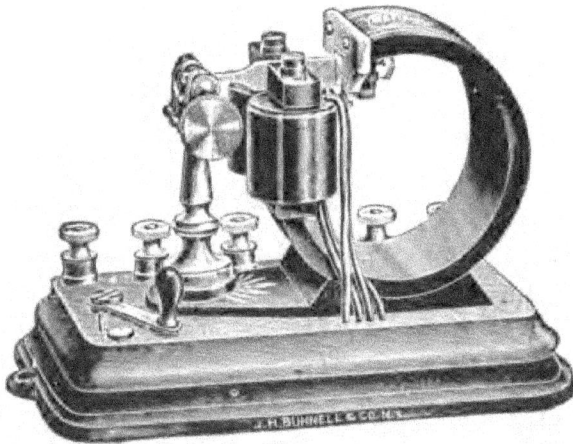

FIG. 119. — Polarized Relay.

In another type the winding is placed upon the soft-iron armature which oscillates between permanent magnets. In the bi-polar telephone receiver the permanent magnet is also the yoke for the electromagnet, and the diaphragm is the armature. In Figs. 120 and 121 the armatures are permanent magnets. These give a general idea of the action of polarized electromagnets. Owing to the fact that like poles repel while unlike poles attract one another, it is readily seen that when the electromagnet is excited, one of its poles will be N and the other S; therefore, the armature will be attracted

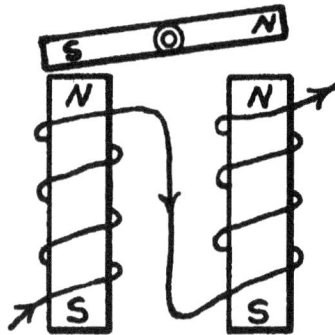

FIG. 120. — Polarized Electromagnet.

on one side and repelled on the other, this action depending upon the polarity of the electromagnet. Hence, the position of the armature is controlled by the direction in which the current flows through the coils.

Polarized electromagnets are very sensitive, respond to alternating currents, and may be worked with great rapidity, — the synchronous action depending upon the inertia of the armature. The

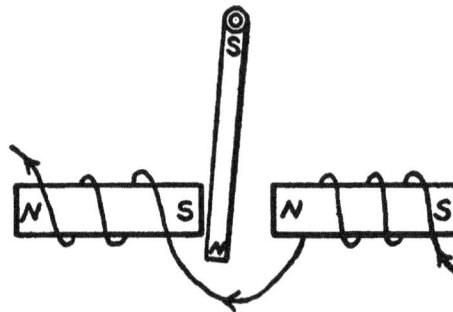

FIG. 121. — Polarized Electromagnet.

armature may also be biased, by means of a spring, for use with direct currents, which action is extremely sensitive. This practice is common in connection with relays used in wireless-telegraph calling apparatus.

When it is desired to balance the armature of a polarized electromagnet so that the armature may be moved in either direction, at will, according to the direction of the current, the device in Fig. 122 may be employed. The field of the permanent magnet tends to normally hold the armature in a balanced position.

The great sensitiveness of polarized electromagnets as compared with those which are non-polarized is because of the greater change in flux density. All other conditions being equal, the attraction is proportional to $\mathcal{B}^2$. Hence, if

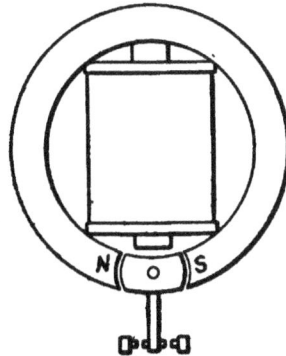

FIG. 122. — Polarized Electromagnet.

under the influence of the permanent magnet alone $\mathcal{B} = 1000$, $\mathcal{B}^2 = 1,000,000$. If now the electromagnet alone produces a flux density $\mathcal{B} = 5$, $\mathcal{B}^2 = 25$. With the complete polarized electromagnet, however, the total flux density after the current flowed would be $\mathcal{B} = 1005$ and $\mathcal{B}^2 = 1005^2 = 1,010,025$, an increase in attraction proportional to $1,010,025 - 1,000,000 = 10,025$. Hence, in this case, the polarized electromagnet would have 401 times as great an attraction for a change of 5 lines per square centimeter as would be obtained with the electromagnet alone. The above results could, however, only be obtained under ideal conditions.

CHAPTER X

ELECTROMAGNETIC PHENOMENA

69. INDUCTION

IF a conductor be passed through a magnetic field at an angle to the lines of force, an e. m. f. will be generated in the conductor. A similar effect may be obtained by varying the intensity of the magnetic field, the conductor remaining stationary. The maximum e. m. f. will be obtained when the conductor is perpendicular to the lines of force, and when the intensity of the magnetic field is suddenly changed from zero to maximum, or vice versa.

This is the principle employed in all dynamo-electric machines and transformers, and the phenomenon is known as *Induction.*

A similar action takes place between two wires arranged side by side, or between two coils of wire placed one over the other, when one of the circuits is energized with a current varying in strength. This is due to the varying flux produced by the varying current cutting the adjacent conductor.

The rule expressing the relative directions of the inducing and induced currents is known as *Lenz's law,* and is as follows : *The currents induced in an electric circuit, by changes of the current in, or of the position of, an adjacent circuit through which a current is flowing, are always in such a direction as by their action on the inducing circuit to oppose the change.*

70. Self-induction

When there is a change in the strength of current in a conductor, the change in flux produced by that current establishes a counter-e. m. f. in the conductor, and this phenomenon is called *Self-induction*. Thus, in a straight conductor, or coil of wire, if the current strength increases, the increasing flux generates a counter-e. m. f. which opposes the increasing e. m. f., which causes the increasing current; whereas, if the current be decreasing in strength, the e. m. f. of self-induction acts in the opposite direction.

The presence of iron in the magnetic circuit greatly increases flux, and when the electric circuit is suddenly interrupted, the e. m. f. of self-induction often becomes very great, producing a large spark at the point of rupture. This principle is taken advantage of in electric ignition apparatus.

The magnetic field acts as a reservoir of magnetic energy which returns to the electric circuit an amount of energy corresponding to the electrical energy required to establish the magnetic field.

The practical unit of self-induction is the *Henry*, and is equal to 10^9 absolute units. The self-induction in henrys of any coil or circuit is numerically equal to the e. m. f. in volts induced by a current in it, changing at the rate of one ampere per second.

The term *Flux-turns* (symbol ϕN) is conveniently given to the product of the total flux in the magnetic circuit into the number of turns in the exciting coil.

Inductance (symbol **L**) is the coefficient of self-induction.

$$L = \frac{\phi N}{I \times 10^8};$$

(102)

whence $$\phi N = \mathbf{L}I \times 10^8. \qquad (103)$$

As an example: if a coil have 100 turns of wire, through which a current of 3 amperes is flowing, the ampere-turns will be

$$IN = 3 \times 100 = 300.$$

If 300 ampere-turns produce 200,000 lines of force, *i.e.* $\phi = 200$ kilogausses, the flux-turns will be

$$\phi N = 100 \times 200,000 = 20,000,000, \text{ or } 2 \times 10^7.$$

Then $$\mathbf{L} = \frac{2 \times 10^7}{3 \times 10^8} = \frac{2}{3 \times 10} = 0.06667 \text{ henry.}$$

If the current of 3 amperes dies out uniformly in one second, then the induced e. m. f. is

$$e = \mathbf{L}\frac{I}{t} = 0.06667 \times 3 = 0.200 \text{ volt.}$$

$\mathbf{L}$ is a constant when there is no iron or other magnetic material in the magnetic circuit. When iron is present, as is nearly always the case in practice, the permeability for different degrees of magnetization must be taken into consideration.

71. Time-constant

The phenomenon of self-induction prevents a current from rising to its maximum value instantly, *i.e.* a certain lapse of time is required before Ohm's law, $I = \dfrac{E}{R}$, holds, unless the effects of self-induction be neutralized.

The time-constant is numerically equal to $\dfrac{\mathbf{L}}{R}$ and is the time required for the current to rise to 0.634 or 63.4 per cent of its Ohm's-law value.

Helmholtz's law expresses the current strength at the end of any short time, t, as follows:

$$I = \frac{E}{R}\left(1 - e^{-\frac{R}{L}t}\right), \qquad (104)$$

wherein $e = 2.7182818$, the base of the Napierian logarithms.

Substituting $\dfrac{L}{R}$ for t in (104),

$$I = \frac{E}{R}\left(1 - e^{-1}\right). \qquad (105)$$

Multiplying and dividing the right-hand member by e, we have

$$I = \frac{E}{R}\left(\frac{e-1}{e}\right). \qquad (106)$$

Hence, $\qquad\qquad I = 0.634\,\dfrac{E}{R}. \qquad (107)$

From the above it is seen that the time-constant may be decreased by decreasing the inductance, or by increasing the resistance. If there was no inductance, but with any value for resistance, the current would reach its Ohm's-law value instantly. On the other hand, if there was no resistance, but with any value for inductance, the current would gradually rise to infinity, the relation between time and current being

$$I = \frac{tE}{L}. \qquad (108)$$

The inductance is sometimes called the *electrical inertia* of the circuit.

As an example, assume $E = 20$; $R = 500$; $L = 10$.

The final value of I will be $\frac{20}{500} = 0.04$ ampere, and the time-constant is $\frac{10}{500} = 0.02$ second; that is, the time required for the current to rise to 0.04 ampere $\times 0.634 = 0.02536$ ampere, will be 0.02 second.

72. Inductance of a Solenoid of Any Number of Layers

Louis Cohen has deduced a formula* which is correct to within one half of one per cent where the length of the solenoid is only twice the diameter, the accuracy increasing as the length increases.

The formula is as follows:

$$L = 4\pi^2 m^2 n \left\{ \frac{2r_m^4 + r_m^2 L^2}{(4r_m^2 + L^2)^{\frac{1}{2}}} - \frac{8r_m^2}{3\pi} \right\} + 8\pi^2 m^2$$

$$\{[(n-1)r_1^2 + (n-2)r_2^2 + \cdots] [\sqrt{r_1^2 + L^2} - \tfrac{7}{8}r_1]$$
$$+ \tfrac{1}{2}[n(n-1)r_1^2 + (n-1)(n-2)r_2^2$$
$$+ (n-2)(n-3)r_3^2 + \cdots]\left[\frac{r_1 d_1}{(r_1^2 + L^2)^{\frac{1}{2}}} - d_1 \right]\}. \quad (109)$$

wherein $n =$ number of layers,
          $r =$ mean radius of the solenoid,
$r_1, r_2, r_3, \cdots r_n =$ mean radii of the various layers,
          $L =$ length of the solenoid,
          $d_1 =$ radial distance between two consecutive layers,
          $m =$ number of turns per unit length.

All the above are expressed in centimeters.

For a long solenoid, where the length is about four times the diameter, only the first two members of equation (109) need be used, i.e. the formula ends with $-\tfrac{7}{8}r_1]$.

* *Electrical World*, Vol. L, 1907, p. 920.

The above is for a solenoid with an air-core.

Maxwell's formula, while not so accurate, is very convenient for rough calculations, and is as follows:

$$L = \tfrac{4}{3}\pi^2 m^4 L(r_e - r_i)(r_e{}^3 - r^3), \qquad (110)$$

wherein r_e and r_i are the external and internal radii of the solenoid.

73. EDDY CURRENTS

Electric and magnetic circuits are always interlinked with one another. The current in a coil of wire surrounding a bar of magnetic material establishes a magnetic field which is greatly augmented by the permeability or multiplying power of the magnetic material.

When a variation in the strength of current in the coil takes place, there will be a corresponding variation (when the core is not saturated) in the strength of the magnetic field of the iron, and since the iron or steel constituting the core is a conductor of electricty, an e. m. f. will be established in it at right angles to the direction of the flux in the core; that is, there will be mutual induction between the varying current in the coil and the iron core.

These induced currents are called *Eddy Currents*, and are largely overcome by subdividing the core in the direction of the flux. By this method the path of the flux is not interfered with, but the electric circuit in the core may be destroyed to a sufficient degree for all practical purposes.

CHAPTER XI

ALTERNATING CURRENTS

74. SINE CURVE

IN an alternating-current generator of the type depicted in Fig. 123, the e. m. f. will vary as the sine of the angle through which the conductor travels through the magnetic field, the rate of travel and the strength of field being uniform.

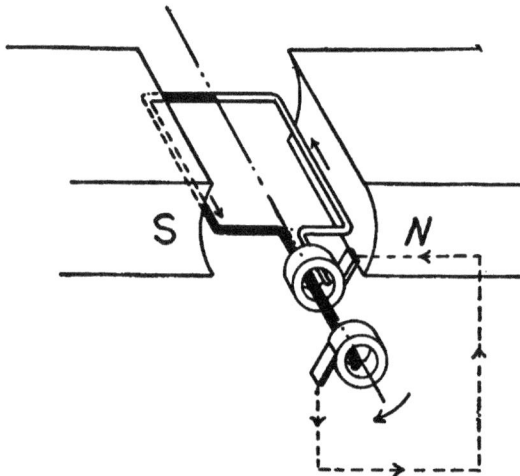

FIG. 123.—Production of Alternating Current.

Referring to Fig. 124, it is evident that the e. m. f. will change from zero, at 0°, to its maximum value at 90°, and will then fall to zero again at 180°. The same operation will be repeated as the conductor revolves from 180° to 360°, but the direction of the e. m. f. will be reversed.

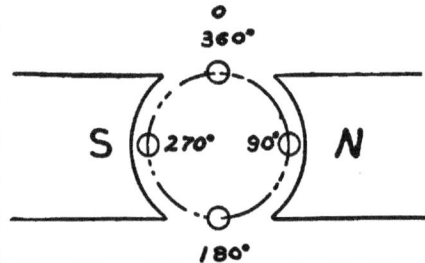

FIG. 124.—Relative Angular Positions of Conductor.

154

The instantaneous values of the e. m. f. may be plotted in the form of a curve, as in Fig. 125. This is known as the *Sine Curve* or *Sinusoid*.

One complete revolution is called a *Cycle* (symbol ∿), and one half of this one *Alternation*. Hence, one cycle consists of two alternations.

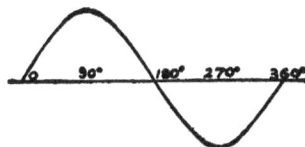

FIG. 125.—Sinusoid.

The *Period* of an alternating current is the time required to complete one cycle. The number of cycles per second is the *Frequency* (symbol f). In this country (U.S.A.) the standard frequencies are 25 ∿ and 60 ∿.

75. PRESSURES

Referring to Fig. 125 it is seen that the e. m. f., during one alternation, is *Maximum* at 90° or 270°, and that the *Average* or *Arithmetical Mean* e. m. f. is proportional to the average ordinate of the curve.

If the maximum ordinate at 90° (or 270°) be considered as 1, then the area of the curve, for one alternation, is 2. The length 0° to 180° = π.

Since the arithmetical mean e. m. f., E_A, is proportional to the mean ordinate,

$$E_A = \frac{2}{\pi} E = 0.637\, E, \qquad (111)$$

for the positive half-wave, and − 0.637 for the negative half-wave. As these two quantities cancel each other, the mean e. m. f. for the whole wave is zero.

The *Effective Pressure*, E_e, is the pressure which will, when applied to a non-inductive resistance, cause a flow of current which produces the same amount of heat as the corresponding current caused by a direct e. m. f. of the same number of volts. That is, an effective pressure of one volt will cause an alternating current of one ampere, effective mean value, through one ohm resistance; and will produce heat at the rate of one watt. *The effective pressure is equal to the square root of the mean of the squares of the successive pressures during one alternation,* or

$$E_e = \sqrt{\frac{0^2 + 1^2}{2}} E = \sqrt{\frac{1}{2}} E = 0.707 \, E. \qquad (112)$$

The squares of negative numbers are positive, as well as those of positive numbers; therefore, the effective mean values are all positive, and the effective pressure is the same for the whole wave as for the half.

From the preceding equation, it follows that

$$E_e = 1.11 \, E_A. \qquad (113)$$

The effective e. m. f. is the e. m. f. referred to in rating alternating-current (A. C.) apparatus, and a common commercial pressure is 104 volts.

From what was said in Art. 70, it is evident that there will also be a counter-e. m. f. due to inductance. This is called the *Electromotive Force of Self-induction* (symbol E_L), and is always opposed to the inducing force. This self-induced pressure is the greatest when the alternating current is the least, and vice versa. That is, it lags 90°, or one quarter of a cycle, behind the current. The current itself lags behind the *Impressed* e. m. f. any amount between 0° and 90°, according to circumstances, but usually in practical apparatus containing inductances purposely introduced (such as transformers and choke coils) nearly 90°; so that the e. m. f. of self-induction lags nearly 180°, or half a cycle, behind

the impressed e. m. f. That is, the e. m. f. of self-induction is very nearly in opposition to the impressed e. m. f.
The two e. m. f.'s are then said to be *out of phase* with
each other. This is shown in Fig. 126.

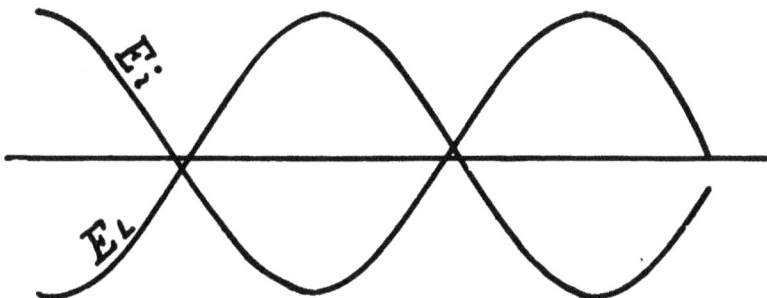

FIG. 126. — Impressed e. m. f. balancing (nearly) e. m. f. of Self-Induction.

The *Impressed Pressure* (symbol E_i), when applied to
a circuit containing both resistance and inductance, is
considered as being split up into two components, one
of which is in opposition to and balances the e. m. f. of
self-induction, and, therefore, leads the impressed e. m. f.
somewhat, and the other, called the *Active Pressure*
(symbol E_a), which causes current to flow, and which is
always in phase therewith and proportional thereto.
The component which balances the e. m. f. of self-induction is called the *Self-induction Pressure* (symbol E_s).
The active pressure is the resultant of the impressed
and self-induction pressures.

Therefore, $E_a = \sqrt{E_i^2 - E_s^2}.$ (114)

Referring to Fig. 127, E_a is the resultant or active
pressure required to send the current I, which is in phase
with E_a, through a given resistance, and E_s is the self-
induction pressure. Curve E_i is the impressed pressure
or applied e. m. f., and its instantaneous values are equal

to the algebraic sums of the instantaneous values of curves E_a and E_s. It is somewhat in advance of I and E_a.

The effective value of the induced e. m. f. may be

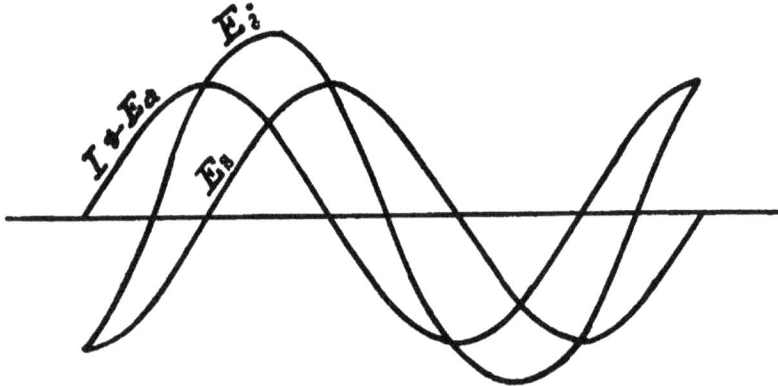

FIG. 127. — Phase Relations when $E_a = E_s$.

readily calculated when the inductance **L** is known. ϕ is the maximum flux, and this is cut four times by the coil during each cycle, since the flux rises from zero to maximum; falls to zero; increases to maximum in the opposite direction, and falls again to zero. Hence, if the coil * has N turns and the frequency is f cycles per second, the average or arithmetical mean e. m. f. will be

$$E_A = \frac{4 \phi N f}{10^8}, \qquad (115)$$

and since $E_e = 1.11\, E_A$ (113),

$$E_e = \frac{4.44\, \phi N f}{10^8}. \qquad (116)$$

If the inductance of a circuit is **L** henrys,

$$\mathbf{L} = \frac{\phi N}{I_m \times 10^8}, \qquad (117)$$

---

* A coil as in Fig. 21, p. 36, is here meant. For inductance due to coil of any dimensions, see Art. 72.

wherein I_m is the maximum current, which will be

$$I_m = I_e\sqrt{2}, \qquad (118)$$

wherein I_e is the effective current.

Therefore, $\qquad I_e = \dfrac{I_m}{\sqrt{2}} = 0.707\ I_m. \qquad (119)$

Hence, $\qquad \mathbf{L} = \dfrac{\phi N}{I_e\sqrt{2} \times 10^8}, \qquad (120)$

and $\qquad \phi N = I_e\sqrt{2}\ \mathbf{L} \times 10^8, \qquad (121)$

wherein $\phi N =$ flux-turns.

But from (116)

$$E_e = \frac{4.44\ \phi N f}{10^8} = \frac{4 \times \dfrac{0.707}{0.636}\ \phi N f}{10^8} = 4 \times \frac{\dfrac{1}{\sqrt{2}}}{\dfrac{2}{\pi}} \times \frac{\phi N f}{10^8}. \quad (122)$$

Substituting the value of ϕN from (121) in (122),

$$E_e = 4 \times \frac{\dfrac{1}{\sqrt{2}}}{\dfrac{2}{\pi}} \times f I_e\sqrt{2}\ \mathbf{L}. \qquad (123)$$

Hence, $\qquad E_e = 2\,\pi f \mathbf{L} I_e. \qquad (124)$

The expression $2\,\pi f$ is called the *Angular Velocity* (symbol ω).

Representing the e. m. f. of self-induction by E_L and the effective current by I,

$$E_L = 2\,\pi f \mathbf{L} I, \qquad (125)$$

or $\qquad E_L = \omega \mathbf{L} I. \qquad (126)$

76. Resistance, Reactance, and Impedance

The expression $2\pi fL$ is the resistance R_L, due to self-induction, and is known as *Inductive Reactance*.

Hence,
$$I = \frac{E_L}{R_L} = \frac{E_L}{2\pi f\mathbf{L}}. \qquad (127)$$

The apparent resistance offered to the impressed e. m. f. is known as *Impedance* (symbol Z), and *is equal to the square root of the sum of the squares of the resistance and reactance*,

or
$$Z = \sqrt{R^2 + 4\pi^2 f^2 \mathbf{L}^2}. \qquad (128)$$

Then,
$$E_i = IZ = I\sqrt{R^2 + 4\pi^2 f^2 \mathbf{L}^2}, \qquad (129)$$

and
$$I = \frac{E_i}{\sqrt{R^2 + 4\pi^2 f^2 \mathbf{L}^2}}. \qquad (130)$$

If $\mathbf{L} = 0$,
$$I = \frac{E_i}{\sqrt{R^2}} = \frac{E}{R}, \qquad (131)$$

which is Ohm's law.

77. Capacity and Impedance

The *Capacity* of an alternating-current circuit is the measure of the amount of electricity held by it when its terminals are at unit difference of potential.

An example of capacity is found in the familiar type of *electrical condenser*, in which sheets of tin-foil are insulated from one another and arranged as in Fig. 128.

Fig. 128.—Condenser.

The effect of capacity is directly opposed to self-induction, and it is possible, by properly adjusting the capacity and inductance of a circuit so that they will neutralize one another, to bring the laws of the alternating current under those of direct.

If E_c be the pressure applied to a condenser and C be its capacity in *Farads*,

$$I = 2\,\pi f\, CE_c, \qquad (132)$$

or

$$I = \omega\, CE_c, \qquad (133)$$

and

$$E_c = \frac{I}{\omega C}. \qquad (134)$$

From (134) it is evident that the resistance due to capacity is

$$R_c = \frac{1}{\omega C}. \qquad (135)$$

The impedance due to resistance and inductance in series was given in equation (128). This may be written

$$Z = \sqrt{R^2 + L^2\omega^2}. \qquad (136)$$

The impedance due to resistance and capacity in series is

$$Z = \sqrt{R^2 + \frac{1}{C^2\omega^2}}, \qquad (137)$$

and for resistance, inductance, and capacity in series

$$Z = \sqrt{R^2 + \left(L\omega - \frac{1}{C\omega}\right)^2}. \qquad (138)$$

78. RESONANCE

When $L\omega = \dfrac{1}{C\omega}$, $Z = R$. This condition is called *Resonance*.* This effect is shown in Fig. 129.

The practical unit of capacity is the *Microfarad* (one millionth of a farad).

* For further particulars see Foster's *Electrical Engineers' Pocket Book*.

For resonance, $$\mathbf{L}\omega = \frac{1,000,000}{\omega C_m},\qquad(139)$$

wherein C_m is the capacity in microfarads.

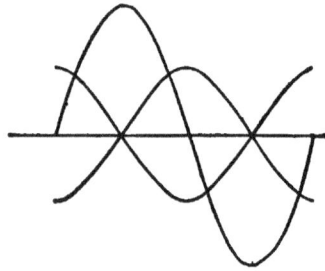

From (139)

$$\omega = 10^3\sqrt{\frac{1}{LC_m}}.\qquad(140)$$

Since

$$\omega = 2\,\pi f,\qquad(141)$$

$$f = \frac{10^3}{2\,\pi}\sqrt{\frac{1}{LC_m}},\qquad(142)$$

FIG. 129. — Conditions for Resonance.

or $$f = 159.2\sqrt{\frac{1}{LC_m}}.\qquad(143)$$

The opposing capacity and inductance e. m. f.'s usually set up local pressures much greater than the impressed pressure.

The e. m. f. at the terminals of an inductance, necessary to force a current through it, is

$$E_L = \omega LI,\qquad(126)$$

and since, for resonance,

$$I = \frac{E}{R},\qquad(131)$$

$$E_L = \frac{E\omega L}{R}.\qquad(144)$$

The e. m. f. necessary to force a current through a capacity is

$$E_c = \frac{I}{\omega C},\qquad(145)$$

or $$E_c = \frac{10^6 I}{\omega C_m}.\qquad(146)$$

Substituting $\dfrac{E}{R}$ for I in (146),

$$E_c = \frac{10^6 E}{R\omega C_m}. \tag{147}$$

As an example, refer to the conditions shown in Fig.

R=6 OHMS

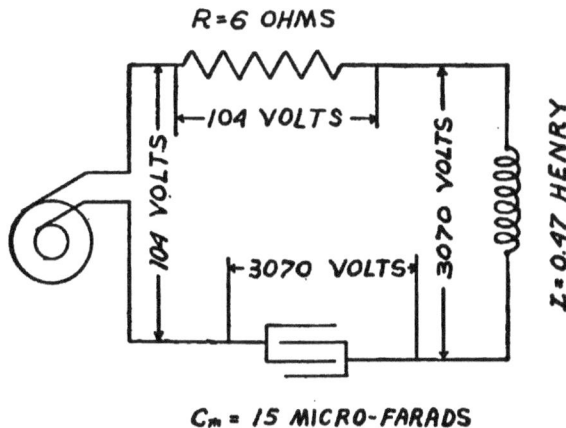

104 VOLTS

104 VOLTS

3070 VOLTS

3070 VOLTS

$L = 0.47$ HENRY

$C_m = 15$ MICRO-FARADS

FIG. 130.—Effects of Resonance.

130. Here $L = 0.47$, $C_m = 15$, $E = 104$, $R = 6$, $f = 60$ cycles, $I = \frac{104}{6} = 17.3$ amperes.

From (144) $\qquad E_L = \dfrac{E\omega L}{R}.$

Since $\qquad\qquad \omega = 2\pi f,$

$$E_L = \frac{104 \times 2\pi \times 60 \times 0.47}{6} = 3070 \text{ volts,}$$

which is the e. m. f. across the terminals of the inductance.

The e. m. f. at the terminals of the capacity is

$$E_c = \frac{10^6 E}{R\omega C_m}. \tag{147}$$

Hence, $E_c = \dfrac{1,000,000 \times 104}{6 \times 2\,\pi \times 60 \times 15} = 3070$ volts.

If the resistance R was 5 ohms, the e. m. f.'s across the terminals of the inductance and condenser would each be 3675 volts for resonance.   Hence, it is seen that the smaller the resistance, the greater will be the local e. m. f.'s.

In practice, so exact values cannot be obtained owing to the fact that the e. m. f. is not, as a rule, a pure sine curve function, as has been assumed in the foregoing. Although complete resonance may not be obtained, in practice, at commercial frequencies, the partial neutralization, due to the placing of capacity and inductance in series, tends to make the local e. m. f. higher than the impressed.

79. Polyphase Systems

When the *angle of lag* between two currents is zero, they are in *phase*.   If the angle of lag is 90°, they are in *quadrature*, and if 180°, they are in *opposition*.

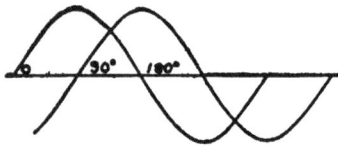

FIG. 131.—Two-phase Currents.

In Fig. 131 are shown two current waves in quadrature. If each of these currents were fed into separate lines, a *two-phase* system would be obtained; the currents differing in phase by 90° or one quarter period.

In three-phase systems the currents differ in phase by 120° (one third period).   This effect is shown in Fig. 132. It is an easily demonstrated

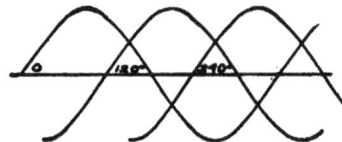

FIG. 132.—Three-phase Currents.

property of these currents that their algebraic sum, at any given instant, is zero, or, in other words, at any given instant, there is one of the three currents which is equal in strength to the sum of the other two currents, and opposite in direction thereto.   Consequently, one wire of each of the three circuits may be dispensed with, and the three currents carried on the three remaining wires, one of which, at any given instant, acts as the return for the other two, or two act as the return for the other one.

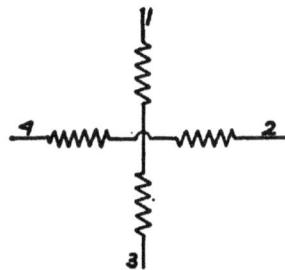

Fig. 133. — Two-phase System.

The general plan of the two-phase system is shown in Fig. 133, while the two common three-phase systems are diagrammatically shown in Figs. 134 and 135,

the former being called the *Star* or Y connection and the latter the *Delta* (Δ) connection.

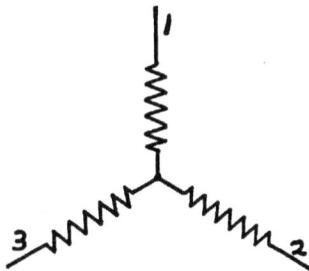

Fig. 134. — Star or Y Connection, Three-phase.

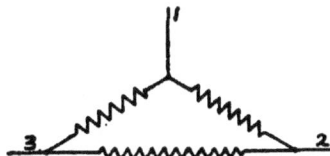

Fig. 135.—Delta Connection, Three-phase.

80. HYSTERESIS

When an alternating current flows through the winding of an electromagnet, the magnetism in the core is rapidly and completely reversed, the magnetizing force rising from zero to maximum; falling to zero; then to its maximum value in the negative direction, and back again to zero.

Theory indicates that the molecules of the magnetic material in the core are reversed with each reversal of the magnetizing force, and that a certain molecular friction takes place which is due to the coercive force in the magnetic material. This friction causes a loss of energy in the form of heat. The phenomenon is known as *Hysteresis*, and the energy loss as the *Hysteresis Loss*.

The *Hysteresis Loop* in Fig. 136 shows the relative values for $\mathscr{H}$ and $\mathscr{B}$ in a soft-iron ring. When the iron was first gradually magnetized, the curve started from the origin, but, owing to the coercive force, the curve can never again pass through the origin after the iron is once magnetized. The hysteresis loss is proportional to the area of the hysteresis loop.

Steinmetz, after exhaustive experiments, has found this loss to be

$$w_c = f n_c \mathscr{B}^{1.6} \times 10^{-7} \tag{148}$$

wherein w_c = watts lost per cubic centimeter of iron,

f = number of complete reversals (cycles) per second,

and n_c = *hysteretic constant*,

which varies with different grades of iron and steel, 0.003 being a good average for thin sheet iron.

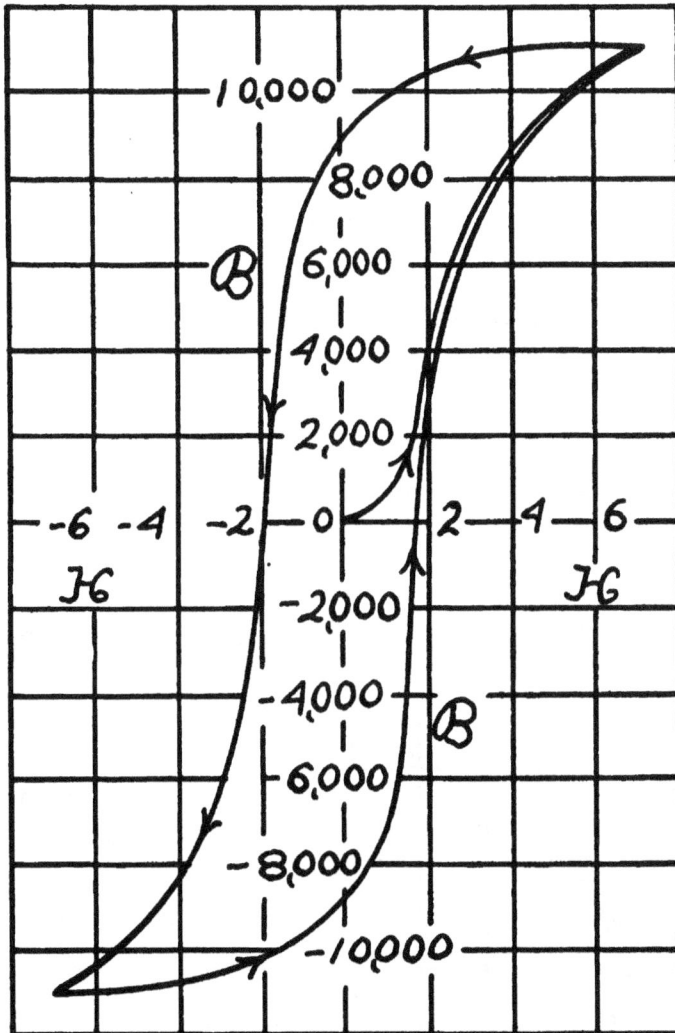

FIG. 136. — Hysteresis Loop.

CHAPTER XII

ALTERNATING-CURRENT ELECTROMAGNETS

81. EFFECT OF INDUCTANCE

IN an alternating-current (A. C.) electromagnet, the inductance will vary with the relative positions of the

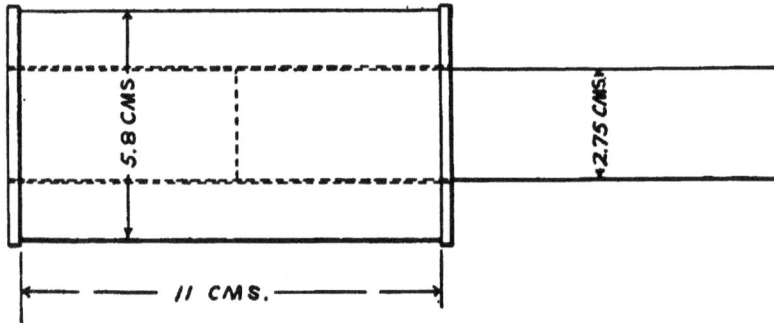

FIG. 137. — A. C. Solenoid.

coil and plunger or armature.

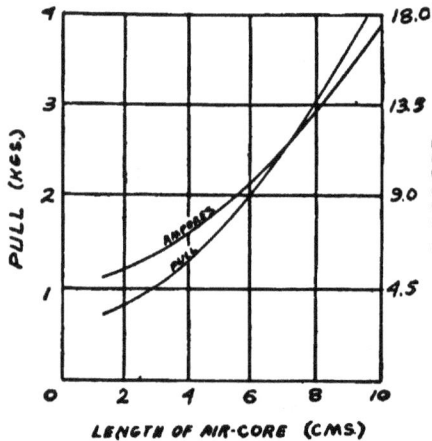

FIG. 138. — Characteristics of A. C. Solenoid.

Hence, the strength of the current will vary also. To illustrate this effect, some tests * made by the author will be cited.

The solenoid in Fig. 137 was tested with a core or plunger consisting of a bundle of soft-iron wires, and Fig. 138 shows a result of a test of this solenoid on a 104-volt, 60-

* *American Electrician*, Vol. XVII, 1905, p. 467.

cycle circuit. It will be noticed that, as the plunger is withdrawn, the current increases, thereby increasing the ampere-turns, and consequently the pull on the plunger.

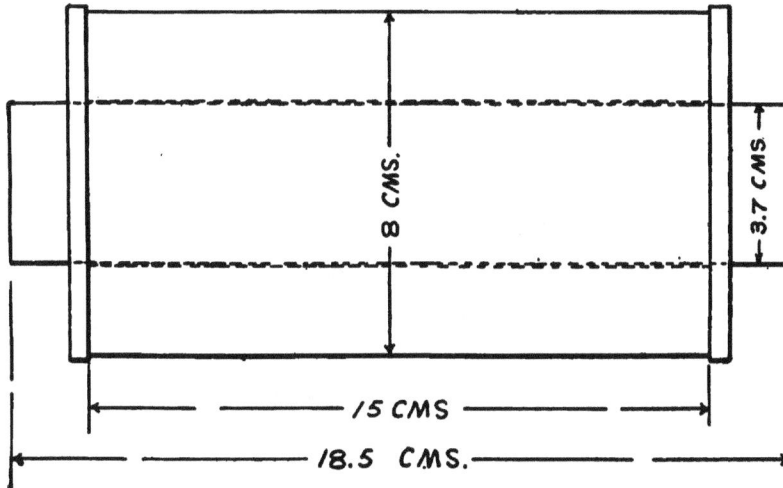

FIG. 139. — Inductance Coil with Taps.

The coil in Fig. 139 has a laminated core and also three taps, making four test windings. The resistances were 0.11, 0.23, 0.36, and 0.63 ohm, respectively, and the turns 131, 261, 388, and 609, respectively. Figure 140 shows the resistance and turns and the corresponding current on 104 volts and 60 cycles. This

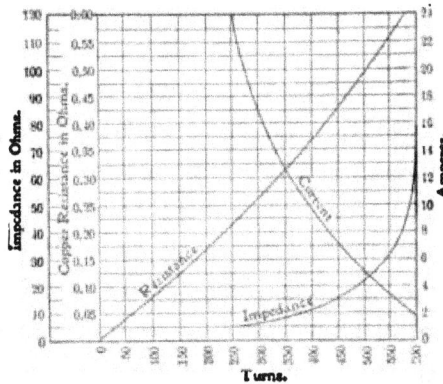

FIG. 140. — Characteristics of Inductance Coil with Taps.

also shows the ratio between resistance and impedance for different numbers of turns.

The curves in Fig. 141 are plotted from a test of the entire winding of 609 turns and 0.63 ohm, and show the effect of inserting different proportions of the total amount of the iron wires constituting the core, which was 19 cm. long and 3.8 cm. in diameter.

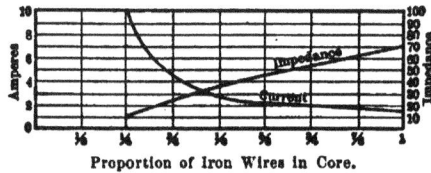

Proportion of Iron Wires in Core.

FIG. 141.—Effect due to Varying Iron in Core.

The test plotted in Fig. 140 shows that while the resistance in the winding was only 0.63 ohm, the total impedance was $\frac{104}{1.5} = 69.4$ ohms, making the resistance of the copper in the winding practically a negligible factor.

82. INDUCTIVE EFFECT OF A. C. ELECTROMAGNET

The constantly changing flux which sets up an e. m. f. of self-induction also tends to induce currents in the cores, frame, and other metallic parts of the magnet. These induced currents oppose the current in the coil and resist any changes in the magnetism. This is in accordance with Lenz's law. (See Art. 69, p. 148.)

The natural method of reducing these induced or secondary currents is to subdivide the core at right angles to the direction of the flux. Thin laminæ, insulated from one another, are employed.

If the spool be of metal, it should be slotted longitudinally with one slot through the tube and washers. This general rule should be followed for all metal parts.

In an A. C. solenoid or plunger electromagnet, the flux tends to pass through the metal (usually brass) tube, in which the plunger travels, at an angle with the direction of travel. Hence, the tube is liable to be heated at the position of the end of the plunger, unless a great many slots are milled, or holes bored, in the tube. This effect is even more marked at the mouth of the plunger electromagnet, where the flux passes from the frame to the plunger.

83. CONSTRUCTION OF A. C. IRON-CLAD SOLENOIDS

In Fig. 142 is shown a modified form of iron-clad solenoid, for the reduction of noise. By this con-

struction, the noise or chattering due to the striking of the plunger against the iron frame at each alternation is somewhat reduced.

In any type of A. C. electromagnet, the plunger, cores, or armature, though laminated, must be solidly constructed so that there can be no lateral vibration of the laminæ, as otherwise humming would result. There is also a tendency on the part of the plun-

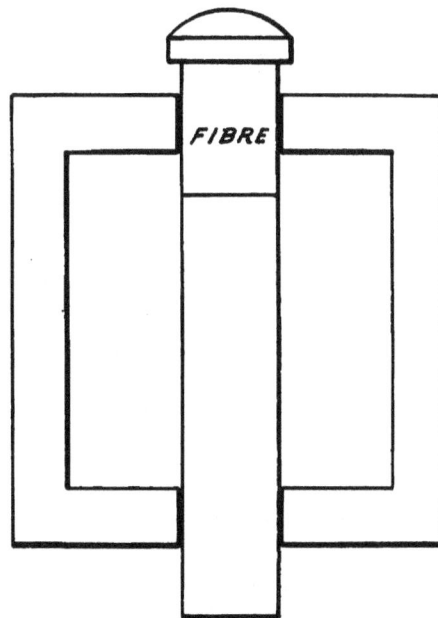

FIG. 142. — Method of Reducing Noise in A. C. Iron-clad Solenoid.

ger to vibrate sidewise, where it passes through the iron frame, which may be avoided by using a guide and making the hole through the frame considerably larger than the core.

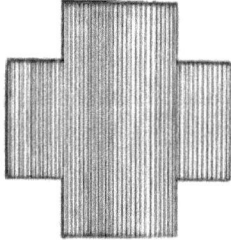

FIG. 143.—Laminated Core.

When the cores or plungers are built up of thin sheet iron, they are usually square in cross-section or of the form shown in Fig. 143, while those made of iron wires are round. The type of core shown in Fig. 143 is for use in a round tube.

84. A. C. PLUNGER ELECTROMAGNETS

The frame and plunger of the single-coil plunger electromagnet, shown in Fig. 144, are constructed of

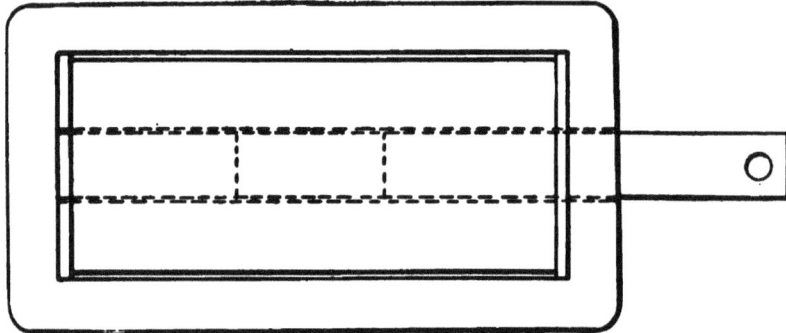

FIG. 144.—A. C. Plunger Electromagnet.

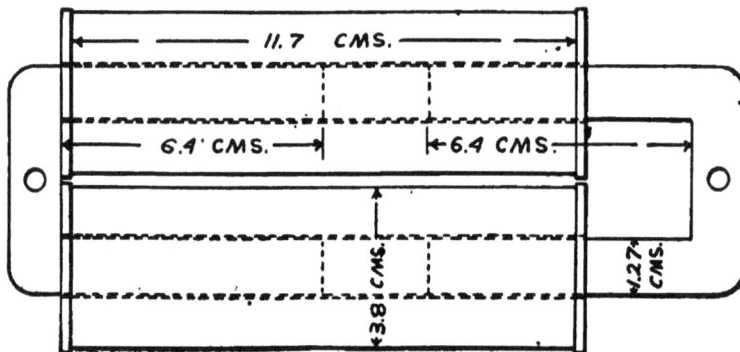

FIG. 145.—Two-coil A. C. Plunger Electromagnet.

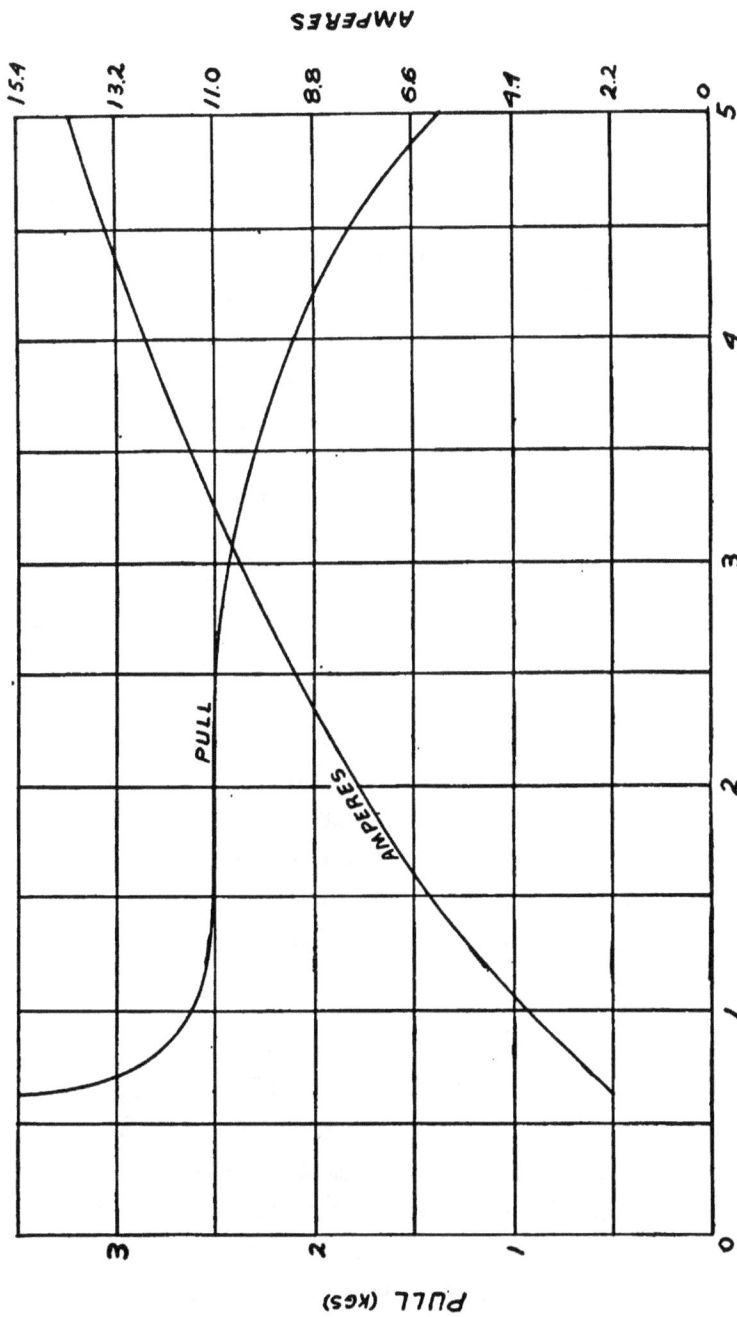

FIG. 146. — Characteristics of Two-coil A. C. Plunger Electromagnet.

thin iron laminæ riveted together. While this is easily constructed after the punches and dies are made, it is rather expensive to make without the above special tools. Hence, where it is intended for intermittent work, the frame often consists of a solid casting.

The frame and plunger of the two-coil plunger electromagnet in Fig. 145 consists of two U-shaped, laminated parts, upon one of which the spools are mounted. This is a simple form of construction. The results of a test* of this magnet by the author on a 104-volt, 60-cycle circuit is shown in Fig. 146.

Each spool was wound with 1400 turns of No. 20 B. & S. wire, and connected in parallel. From Fig. 146 it is seen that while the current is very strong at the beginning of the stroke, it falls to a low value after the magnet performs its work. This is a decided advantage.

This magnet is capable of a much stronger pull, with correspondingly stronger current, but it was designed for nearly continuous service, and, therefore, would overheat if the impedance were made lower.

For maximum efficiency, the center of the air-gap should be at the center of the coil, as in direct-current electromagnets.

85. Horseshoe Type

This magnet, illustrated in Fig. 147, is easily made from the U-shaped laminæ described in Art. 84. In the design of these cores, great care must be exercised in the selection of the proper wire or laminæ, for if the wire or laminæ be too large in cross-section, the

* *American Electrician*, Vol. XVII, 1905, pp. 467–468.

loss due to eddy currents will be too great; on the other hand, if the wires be too small in cross-section, or the insulation between them be too thick, the magnetic reluctance will be so great as to more than offset the evil effects of the eddy currents. The spools are, of course, slotted, if of metal.

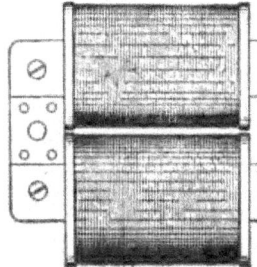

Fig. 147.—A. C. Horse-shoe Electromagnet.

Whenever it is feasible, from a mechanical standpoint, to use spools of insulating material, it is electrically advantageous to do so, as the induced currents in the spools will be eliminated.

86. A. C. Electromagnet Calculations

From (116)
$$E = \frac{4.44\phi Nf}{10^8},$$

wherein E is the impressed e. m. f., ϕ the total flux, N the number of turns in the coil, and f the frequency.

Since $\phi = \mathcal{B}A$,

$$E = \frac{4.44\,\mathcal{B}ANf}{10^8}. \tag{149}$$

On account of the heating due to hysteresis and eddy currents, A. C. electromagnets are usually worked at lower flux densities than for D. C. magnets.

The exact value of the current cannot be easily calculated, due to the variable induction in the iron, but if a curve be plotted showing the magnetic flux for each instantaneous current strength, an accurate value of the effective current may be obtained.

If the saturation curve is considered to be a straight *

* D. L. Lindquist, *Electrical World*, Vol. XLVII, 1906, p. 1296.

line (which is nearly correct for a long air-gap), and the current at the begining of the stroke is I, then

$$I = c_1 \frac{\mathscr{B}}{N}, \qquad (150)$$

wherein c_1 is a constant.

From equations (149) and (150)

$$EI = \frac{4.44 \, \mathscr{B}ANf}{10^8} \cdot c_1 \frac{\mathscr{B}}{N}, \qquad (151)$$

or

$$EI = \frac{4.44 \, c_1 f A \mathscr{B}^2}{10^8}. \qquad (152)$$

From equation (68) $P = \dfrac{\mathscr{B}^2 A}{8 \, \pi \times 981,000}.$

Transposing, $\mathscr{B}^2 A = 8 \, \pi P \times 981,000.$ (153)

Substituting the value of $\mathscr{B}^2 A$ from (153) in (152),

$$EI = \frac{4.44 \, c_1 f P \times 8 \, \pi \times 981,000}{10^8} = 1.095 \, c_1 f P. \quad (154)$$

If $1.095 \, c_1 = c_2$, then

$$EI = c_2 f P. \qquad (155)$$

From (154) $P = \dfrac{EI}{c_2 f},$ (156)

which shows that the pull decreases as the frequency increases. The efficiency of the magnet also varies with the frequency.

87. POLYPHASE ELECTROMAGNETS

Single-phase electromagnets may be operated on polyphase circuits by connecting the magnet in one of the phases only, or magnets corresponding in number to the number of phases may be connected in the respective phases, with their armatures rigidly connected to a common bar or plate, as in Fig. 148.

D. L. Lindquist * has published the results of tests of polyphase magnets, and has treated the matter

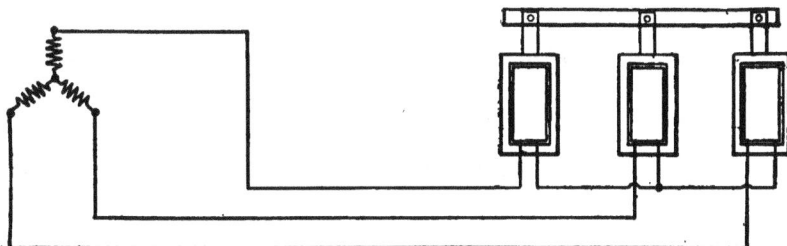

Fig. 148. — Single-phase Magnets on Three-phase Circuit.

thoroughly. The following is abstracted from his articles.

Figure 149 shows a two-phase magnet which consists of two cores practically alike. Each core is built up of a brass spider, b, on

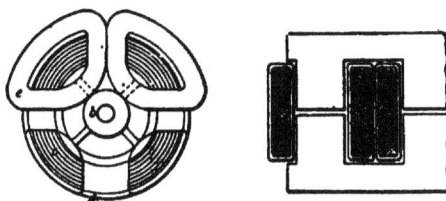

Fig. 149. — Polyphase Electromagnet.

which is wound a spiral of iron band (or ribbon), c; between consecutive layers of iron is a thin sheet of paper fastened with shellac. The interconnection of the four coils of the magnet is shown in Fig. 150.

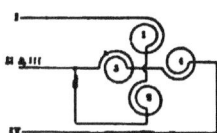

Fig. 150. — Connections of Coils of Polyphase Electromagnet.

Assume now that the two-phase e. m. f.'s impressed upon the core are in time quadrature with each other, and that the e. m. f. waves are of sine shape. Let the instantaneous density in cores 1 and 2 be represented by $\mathcal{B}_a$ and that in cores 3 and 4 by $\mathcal{B}_b$. If the coil resistance and the magnetic leakage are negligible,

* *Electrical World*, Vol. XLVIII, 1906, pp. 128-130 and 564-567.

$$\mathscr{B}_a = K \sin \omega t, \text{ where } K \text{ is a constant,} \quad (157)$$

and $\qquad \mathscr{B}_b = K \sin\left(\omega t + \dfrac{\pi}{2}\right).$ $\qquad\qquad$ (158)

$$\mathscr{B}_b = K \cos \omega t. \qquad\qquad\qquad (159)$$

The total pull is proportional to

$$\mathscr{B}_a{}^2 + \mathscr{B}_b{}^2 = K^2 \sin^2 \omega t + K^2 \cos^2 \omega t = K^2. \quad (160)$$

Consequently, the pull is proved to be constant at any time and equal to the maximum in any one core.

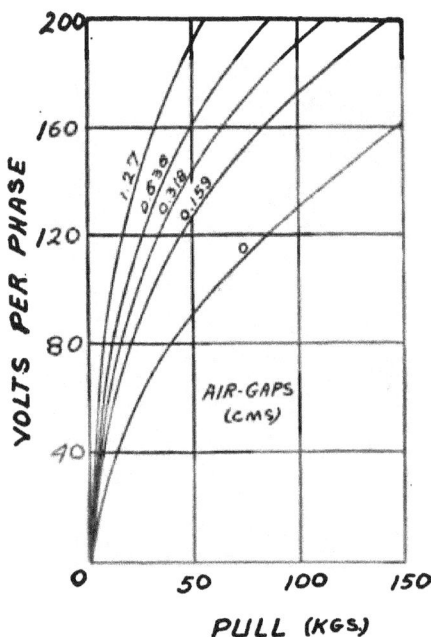

FIG. 151. — Two-phase Electromagnet supplied with Two-phase Current.

As a result of the construction the resultant pull is always exerted through the center axis of the magnet, thus preventing rocking and the consequent chattering.

That a three-phase magnet having three pairs of poles also gives a constant pull can be similarly proved. In practice, however, the two-phase magnet with two pairs of poles has been found suitable for all phases, although it gives slightly less pull when used on three-phase, especially with small air-gaps, as indicated in Figs. 151 and 152, which show tests at 60 cycles on a certain magnet wound with four coils, each containing 220 turns No. 14 wire, the cross-sec-

tional area of the core being 12.5 sq. cm. For large
air-gaps the pull is practically the same in the two cases.

As previously proved,
where the coil resistance
is negligible, and the
magnet has a pair of
poles for each phase, a
polyphase magnet, when
energized by a sine-
shaped e. m. f., exerts a
constant pull. As a
matter of fact, however,
in almost every case the
e. m. f. is more or less
distorted, due to many
causes, the resistance
having a certain in-
fluence, and there must
be some variation in the
pull. If a two-phase

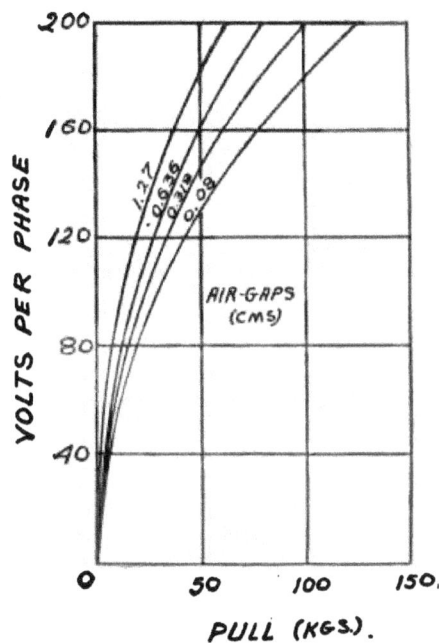

FIG. 152. — Two-phase Electromagnet
supplied with Three-phase Current.

magnet is used on a three-phase circuit, there will be an
additional variation due to this fact. A polyphase magnet should, therefore, never be loaded to such an extent
that the load exceeds the minimum instantaneous pull
of the magnet.

Suppose that the load is in excess of this minimum,
then the conditions would be the same as with a single-
phase magnet; the armature would leave the fixed
pole when the pull was less than the load, causing a
blow when returning. The total pull is proportional to

$$e^2_4 + e^2_2 = \sin^2 a + \sin^2\left(a + \frac{2\pi}{3}\right). \qquad (161)$$

Hence, if the average pull is 1, the maximum pull is 1.5, and the minimum 0.5.

Figure 153 shows the results of tests on the two-

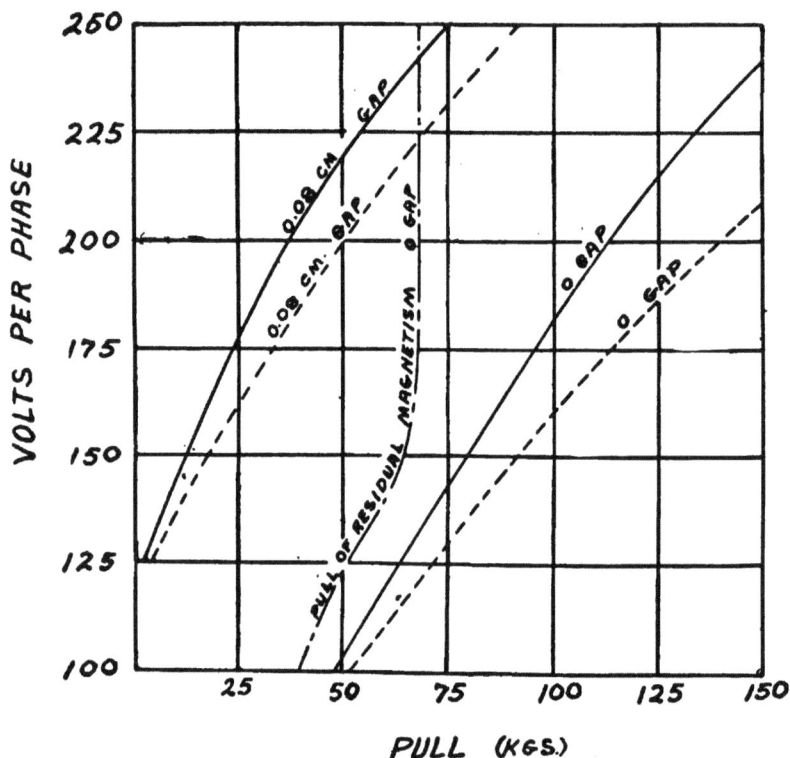

FIG. 153.—Test of Two-phase Electromagnet with Three-phase Current.

phase magnet when energized with three-phase current, the load being increased until the magnet made a noise. As seen from these curves compared with Figs. 151 and 152, giving the pull when energized with two-phase and three-phase current, the magnet will commence to be noisy at about one half the load, but in general it will hold without noise all the load it can lift, so long as the length of motion is not too short. The full lines indicate the pull at which the magnet begins to hum at various voltages with different air-gaps, while the dotted lines show the pull at which chattering begins.

The tests above referred to were made at room temperature, which was approximately 25° C. As the temperature increases the eddy currents decrease, and both the current and the total losses decrease considerably, especially for zero air-gap. A certain test was made to find out how the losses and current consumption varied at different temperatures. A magnet was energized with high voltage to heat the coils and the core. The temperature rise of the core was 57° C. and of coils 69° C.

It was then found that with the same voltage and zero air-gap, the current consumption was only 85 per cent of the current consumption when the magnet was at the room temperature. The ohmic resistance of the coils increased 28 per cent, and the $I^2 R$ losses in the coils were about 92 per cent of the losses when the coils were at room temperature. The total losses were 85 per cent of the losses when the magnet was at room temperature. As the $I^2 R$ losses in the coils were only about 35 per cent of the total losses, the iron losses when hot were only 81.5 per cent of the iron losses at room temperature, due to decrease in eddy currents. This decreasing of the losses when the temperature increases is naturally very advantageous, especially for magnets having to hold their loads continuously.

The fact that the pull is practically independent of the coil resistance, as long as this resistance is fairly well proportioned, is of very great advantage for several reasons. When winding the coils for a magnet of this kind to give a certain pull, no definite size of wire is necessary — merely the right number of turns — and furthermore the temperature of the coil has no influence on the pull. Of course, the coil resistance can be increased to such a value that it has a great deal of influence on the pull, but then the coil is entirely out of proportion, and there is no necessity of using as small a wire as with direct-current magnets because only a small number of turns is necessary.

In general neither resistance nor inductance of a fixed amount can be used for regulating the voltage on an alternating-current magnet. If inductance or resistance is used for regulating the voltage, it is used in conjunction with a switch for inserting more inductance or resistance after the magnet has lifted.

It is impossible to make a single-coil magnet with constant pull, but with the aid of two external resistances a two-phase magnet can be arranged to give constant pull when energized with single-phase current.

Figure 154 gives the connection diagram, while Fig. 155 shows the voltage diagram for this case. All coils

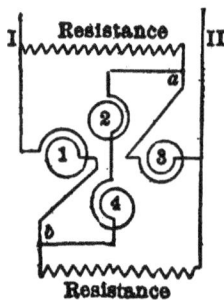

Fig. 154. —Connection Diagram for Polyphase Electromagnet on Single-phase Circuit.

are wound with the same number of turns, and in order to obtain constant pull, it is necessary that all coils be energized alike or that the voltage across every coil be the same, and also that the e. m. f.'s in coils 1 and 3 are in quadrature to the e. m. f.'s in coils 2 and 4. The current through coils 1 and 3 must naturally be considerably larger than that through coils 2 and 4 in order to get the proper phase relation. The poles of

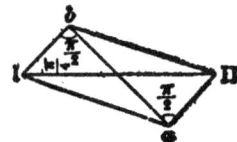

Fig. 155. —Phase Relations in Polyphase Electromagnet on Single-phase Circuit.

coils 1 and 3 are, therefore, made shorter in order to have a certain amount of air-gap between them when the plunger is in the up position.

In order to obtain the required starting pull (with the plunger in the lower position or with the maximum air-gap) only a small amount of resistance is used, and the proper resistance for holding is introduced after the magnet reaches its final position.

CHAPTER XIII

QUICK-ACTING ELECTROMAGNETS AND METHODS OF REDUCING SPARKING

88. RAPID ACTION

It has been shown that by increasing the number of turns in a direct-current magnet, the inductance is increased, which, in turn, increases the time of energizing, and also the time of deënergizing. The induced currents in the coiled core and yoke also have similar effects. Hence, where rapid action is desired, the iron and other metal parts should be subdivided, as in the case of alternating-current magnets.

As the time constant of two coils connected in parallel is only one fourth of what it would be were they connected in series, this method of connection is desirable for rapid-acting magnets.

89. SLOW ACTION

On the other hand, slow action is sometimes desirable. This is, of course, obtained by leaving the cores and yoke solid, and by winding the wire upon a heavy solid brass or copper spool. When the spool consists of insulating material, the retarding effect may be increased by either placing a brass or copper sleeve over the core, or by the use of a short-circuited winding which is separate and distinct from the regular winding. This short-circuited winding may be provided with taps, by means of which the retarding effect may be varied.

Figure 156 is the result of a test * of such a magnet.

* D. L. Lindquist, *Electrical World*, Vol. XLVII, 1906, p. 1295.

90. METHODS OF REDUCING SPARKING

When an electromagnet is connected in a circuit, the phenomenon of inductance tends, upon rupturing the circuit, to increase the total e. m. f., thus producing an abnormal flow of current momentarily. This principle is in common use in electric ignition apparatus, and explains why so large a spark cannot be obtained when a very short contact is made, as may be obtained with a longer duration of contact.

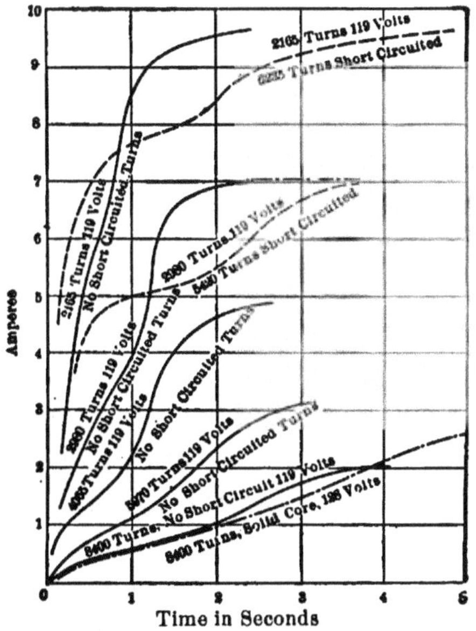

FIG. 156 — Retardation Test of Direct-current Electromagnet.

In the practical application of the electromagnet this sparking is very detrimental, as the repeated sparking at the point of rupture rapidly destroys the contacts.

There are several ways of reducing the sparking. In one method, two exactly similar insulated wires are wound in parallel instead of one, as is customary, thus forming two complete and distinct windings thoroughly insulated from each other, but lying adjacent to each other in every turn of the winding. The two terminals of one winding are then connected together, thus short-circuiting the winding upon itself. The other winding is used as the regular exciting coil.

If now a current be suddenly passed through the exciting winding, a current will also be set up in the short-circuited winding. In this case the effect is electrostatic as well as electromagnetic, since the two windings lie adjacent throughout their entire length ; therefore, there will be no extra sparking at the point of rupture.

In order to obtain this result, however, it is necessary to sacrifice one half the total winding space. Hence in some cases a condenser is connected across the point of rupture, which condenser should have sufficient capacity to absorb all of the extra current due to inductance. By this latter arrangement, all of the winding space may be utilized. In this type, the condenser is sometimes placed around the outside of the winding in order to make the whole magnet compact and self-contained.

Sometimes the condenser is placed across the terminals of the electromagnet. By this arrangement, with proper capacity in the condenser, the sparking due to the inductance is entirely eliminated.

In neither case does the condenser prevent the retarding action of the coil, as in the former case the condenser is short-circuited, and in the latter it is not materially affected by the current at "make." In the former case the contacts are subject to much pitting, due to the short-circuiting of the condenser at "make."

The "break" may be shunted by a resistance which is usually from 40 to 60 times the resistance of the winding of the electromagnet, according to the conditions under which the magnet is to be used.

The electromagnet itself is also often shunted by a high resistance — usually a rod of graphite — which should have about 20 times the resistance of the coil. In any event, the resistance for this purpose must be non-inductive.

The "break" may be shunted by a resistance in series with a battery or other source of energy which will have just sufficient e. m. f. to balance the e. m. f. of the working circuit across the break, but which will provide a path for the extra current at the high potential. See Fig. 157.

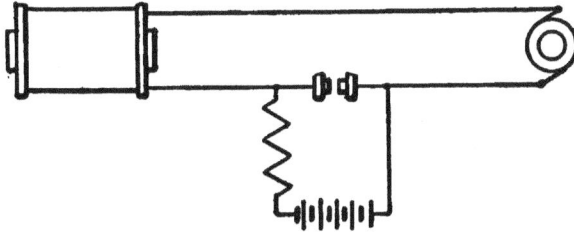

FIG. 157. — Resistance and E. M. F. in Series, in Shunt with " Break."

The winding may also be short-circuited instead of opening the electric circuit, as then the extra current is absorbed in a closed circuit, and there will be no sparking when the shunt is switched out, as there will at that instant be no current in the winding. This is not a very economical arrangement, however, and it is obvious that a very serious short-circuit of the battery or generator would occur when the winding of the electromagnet was short-circuited, unless an external resistance was provided which should remain in circuit after the winding of the electromagnet was short-circuited.

A method very similar to that first described is known as the differential method, in which the windings of the electromagnet are arranged differentially, as in Fig. 158. When the switch is open, the current passes through but one coil, which action magnetizes the core. When the switch is closed, however, the current flows through both coils, in opposite directions, thereby completely neutralizing each other.

Another method is to connect the ends of each layer to common terminals, one at each end of the coil, as in Fig. 203, p. 279.

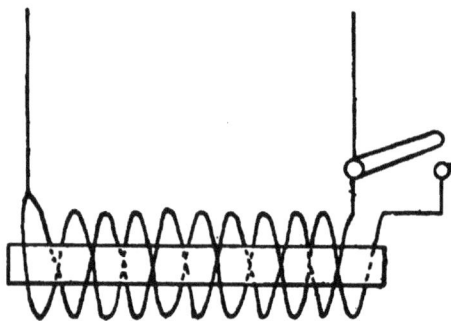

FIG. 158. — Differential Method.

The time constants of the separate circuits being different, owing to the varying diameters of the layers which makes the coefficient of self-induction less and the resistance greater in the outer layers, and vice versa in the inner layers, the extra current flows out at different times for different coils.

Copper sleeves are also sometimes placed over the cores of electromagnets, currents being set up in the sleeves at the time of breaking the circuit, by the lines of force passing through them. It is evident that there will not be so great an inductive effect in the winding when much of the energy is absorbed by the copper sleeve.

Tinfoil is also interposed between the layers of the winding, for the same purpose as above.

Professor Silvanus P. Thompson, who made a comparison test of the following methods, found that the differential method was the best; the multiple-wire winding, tinfoil, and copper sleeve arrangements following in merit in the order given.

The multiple-wire winding referred to above is, in practice, really a multiple-coil winding. This is treated in Art. 163, p. 277.

The spark may be destroyed by a blast of air or by means of a magnet. In the latter case the field of the magnet repels the field established by the arc, thus destroying it.

In general, it may be stated that what is gained in quickness of action is lost in current consumption, and vice versa.

91. METHODS OF PREVENTING STICKING

If the armature of an ordinary horseshoe electromagnet be placed in close contact with the pole pieces before the magnet is energized, it may be easily removed. However, if a direct current be passed through the windings and the magnet be again deënergized, the armature will still be firmly attracted to the pole pieces. This is due to the residual magnetization of the iron. If, however, the iron or ferric portion of the magnetic circuit be broken, so as to introduce a high reluctance, the greater part of the residual charge will disappear.

As this feature is very undesirable, in most electromagnets, non-magnetic stops are usually provided which prevent the armature from actually closing the magnetic circuit, thereby keeping the reluctance so high that the residual magnetization will not have sufficient effect upon the armature as to interfere with the proper operation.

On large electromagnets, in particular, brass or copper pins are forced into the cores to prevent "sticking" of the armature. Where these pins are subject to a heavy blow from the armature, they should have sufficient area to withstand the blow without flattening.

Another method is to place a strip of non-magnetic material over the ends of the cores. This is sometimes made in the form of a cap. On small electromagnets either the ends of the cores or armature, or both, are copper plated, the thickness of the copper being sufficient to prevent sticking. The copper plating also has the property of protecting the iron from oxidation.

In the design of electromagnets, the space occupied by the non-magnetic stops must be taken into consideration.

Merchant Books

CHAPTER XIV

MATERIALS, BOBBINS, AND TERMINALS

92. Ferric Materials

The materials generally used in the construction of
the cores and frame are iron and steel. The best iron
is wrought and Swedish iron. Frames may be made
of cast iron where the reluctance of the air-gap is great,
but the cores should always be made of the best grades
of iron and steel. Cast steel is largely used in the construction of large electromagnets, and tests about the
same as wrought iron (see p. 35).

The magnetic properties of iron and steel depend
largely upon the percentages of carbon in their composition, and also phosphorus, sulphur, manganese, and
silicon.

Wrought iron contains a small percentage of carbon,
and is comparatively soft, maleable, and ductile. It has
a high permeability, but has the disadvantage of being
expensive, unless its form is very simple.

Swedish iron has about the same permeability as
wrought iron.

Cast steel has a very small percentage of combined
carbon and no free carbon, and the best grades do not
contain more than 0.25 per cent of carbon. It is
cheaper than wrought iron, but is hard to obtain on
short notice and in small quantities.

Cast iron is hard and quite brittle, and contains considerable carbon in the free state. It can easily be obtained in almost any desired shape at a low cost.

Irons containing more than 0.8 per cent of combined carbon are of a low magnetic permeability, and those having less than 0.3 per cent are of a high permeability. The combined carbon should be kept as low as possible, while the free carbon may vary from 2 to 3 per cent without having any appreciable effect on the permeability.

In exact work, the permeability is obtained for a sample of each lot of iron or steel used.

93. ANNEALING

The permeability of iron is increased by *annealing*. This is done by heating the iron to a cherry-red, and then allowing it to cool gradually. Special charcoal ovens are provided for this purpose.

During the process of annealing the air must not come into contact with the iron, or oxidation will result. Oxide or rapid cooling makes the iron bad as an electromagnet core, either by scale, which makes the magnet residual on the outside, or by hardening, which makes it residual on the inside, and in the latter case the permeability will be lower than as though the iron were soft.

94. HARD RUBBER

Hard or vulcanized rubber is used extensively in the manufacture of heads or washers of electromagnet spools for telegraph and various other types of electromagnetic apparatus.

It is very brittle at the normal temperature of the air, but becomes soft and pliable when subjected to slight degrees of heat.

Its insulating qualities are excellent, and it is very useful within certain limits of temperature. Its cost is high as compared with fiber, and it has to be handled very carefully in machining. On account of its brittleness it must be softened by heating before forcing on to cores. This is preferably done by placing it in warm water. It may be bent into almost any shape when heated, and will retain its form after becoming cold.

One great advantage of hard rubber is that it may be molded or cast into almost any desired form. It also takes a very high polish, and is therefore very much used where appearance and finish are desirable. It is furnished in both the sheet and rod.

95. VULCANIZED FIBER

The commercial fibers are of three kinds and are sold under the following trade names: Gray Fiber, Red Fiber, and Black Fiber.

The cheaper, and consequently the more common grades, are red and black. Gray fiber is the best, but is somewhat more expensive.

Fiber serves very well as insulating material for low voltages, and it machines quite well, though not so well as hard rubber. However, it is not so brittle as hard rubber.

Fiber readily absorbs moisture which renders it practically useless for high voltages. Nevertheless, it is used extensively for heads of bobbins, etc., and makes an excellent body upon which to place the high-grade insulating materials, such as oiled paper, oiled linen, mica, Micanite, etc.

When used in conjunction with some high-grade insulating material, as just mentioned, fiber is superior to rubber, and does not soften or melt like rubber at comparatively low temperatures, but it is liable to warp on account of its hygroscopic properties.

Fiber is especially adaptable for the making of bobbins, as the tube or barrel is readily made to any size by rolling a thin sheet of it around a mandrel, cementing it together with shellac as it is rolled.

Small heads or washers are usually punched from the sheet. On account of its toughness, fiber is almost exclusively used in making heads for telephone ringers, relays, drops, etc., as the heads may be forced on with great pressure without cracking, thereby insuring firm and solid containing-walls for the winding.

All grades of fiber on the market are not the same, but the best grades show no signs of being built up in layers, and will not readily split with the grain, *i.e.* lengthwise.

Vulcanized fiber is furnished in both the sheet and rod.

96. Forms of Bobbins

Bobbins for electromagnet and solenoid windings are made of various materials and from numerous designs. The natural material for a bobbin of this character is an insulating substance, and as high-grade insulating materials are more expensive than materials of lesser insulating properties, the insulating material in the bobbin usually depends upon the voltage between the winding and the core, or the outer portion of the bobbin. Thus in apparatus where low voltages are to be used the quality of the insulating material used in the bobbin need not be very high.

Next to the insulating properties of the bobbin, the

strength of the material must be considered.   This may again be subdivided into the necessary thickness of the insulating material, for if the insulation be very thick, for a limited size of bobbin, the internal dimensions may be too small for the necessary winding.

Another feature to be considered is the finish of the bobbin which, while often of little consequence where the bobbin is concealed, is of the utmost importance in highly finished instruments where the finish of the bobbin must conform with the rest of the apparatus.

In this case, however, the actual insulating properties of the material of the bobbin may or may not be of great importance owing to the frequent use of low voltage.   In highly finished apparatus the bobbin is usually made of hard rubber which, while an excellent insulator, takes a very high polish.

Bobbins may be generally classified into those with iron cores, and those without iron cores.   Bobbins with iron cores are usually made as shown in Fig. 159.   In this type the heads or washers are forced on to the core, and form the retaining walls of the winding. The core is insulated with paper for low voltage, or with micanite or oiled linen for high voltages.

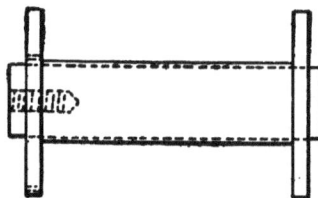

FIG. 159. — Bobbin with Iron Core.

For high voltages, however, special precautions must be taken, which will be discussed farther on.   Bobbins of the type shown in Fig. 159 are sometimes provided with metal washers.   In such cases insulating washers must be placed upon the core between the ends of the winding and the metal washers.   Here the thickness of the insulating washers depends merely upon their insulating properties, as the metal washers take the mechanical strain.

Fiber washers may be forced on to the cores, but rubber will crack unless great care is exercised. A good way to prevent the washers from turning on the core is to put a straight knurl on each end of the core before forcing on the washers. Rubber washers should be dipped in hot water, and then forced on before they become brittle.

Bobbins without iron cores may also be classified into those having metal tubes and those having tubes of insulating materials. In the former type, brass is commonly used for the tube, particularly in solenoids of the coil-and-plunger type where there will be considerable wear on the inside of the tube. In this case the tube must be well insulated according to the voltage used. Where fiber washers are used, a forced fit is usually sufficient. With brass washers, however, it is best to thread the tube and washers and solder them also. All soldering should be done without the use of acid.

Another method, and particularly where a thin brass tube and brass washers are to be used, is to spin up the ends of the tube. Bobbins of this type with thin brass tubes are for use with plunger electromagnets, etc., where they are simply placed over more substantial brass tubes or directly upon the cores themselves.

Bobbins with heavy brass tubes are also assembled by turning off a portion at the ends of the tube, leaving a shoulder which acts as a distance-piece between the washers, and also permits the ends to be spun up. If of metal, the washers may also be soldered.

When a brass tube is used in a bobbin for alternating-current or quick-acting magnets, the tube and washers must be slotted.   Metal bobbins are sometimes cast in one piece.

Bobbins consisting entirely of insulating materials are also vulcanized in one piece with such materials as hard rubber or Vulcabeston.   They are also turned from the solid stock, but this is rather an expensive method. The usual method is to make a tube of paper, rubber, micanite, or other insulating material, and cement the washers thereto with shellac or some other insulating compound.

97. TERMINALS

The terminals of electromagnetic windings may be divided into two classes : (a) those consisting of wires or flexible conductors, and (b) devices to which wires constituting the external circuit may be connected by soldering or by means of screws.

Of the former type, the most natural terminals would be the ends of the wires constituting the winding. While this may be satisfactory where comparatively large wires are used, and generally where there is not much danger of the wire becoming broken, it is usually desirable to employ flexible stranded terminals of copper, thoroughly insulated, whose total cross-section shall be at least as great as the cross-section of the wire in the winding.

In the case of multiple-coil windings, the cross-section of the terminal conductor should at least equal the cross-section of the wire in the winding, multiplied by the number of coils constituting the total winding.

For small coils, with fine wire, a terminal conductor consisting of ten stranded copper wires insulated with a thin coating of soft rubber and covered with silk is very good. The wire should be tinned wherever rubber is used. For larger coils, ordinary lamp cord is excellent.

The inside terminal is the one which is naturally the more important; for if this should become broken, it might be necessary to remove the entire winding, and rewind the coil.

When winding on an ordinary bobbin, the inside terminal should be thoroughly soldered to the end of the wire constituting the winding, before the winding operation is begun, wrapping the terminal proper around the core three or four times in order to take the strain from the wire.

The joint should be thoroughly insulated with a tough insulating cloth or paper, for, unless precautionary measures be taken, the coil is liable to "break down" between the joint and the succeeding layer of wire.

The outside terminal should be connected in substantially the same manner.

When it is necessary to bring the inner terminal to the outside of the coil, strips of mica, micanite, or oiled linen should be placed between the terminal and the rest of the coil to prevent a "break-down." Thin strips or ribbons of copper or brass may be used where the space is limited. In any case the insulating strip should be of ample width, the wider the better.

FIG. 160. — Terminal Conductor.

In cases where binding-screws are fastened to the outside of the coil, both the inner and outer terminal wires should be thoroughly insulated.

For ordinary purposes, where the coil is not exposed to oil or moisture, the type of terminal conductor shown in Fig. 160 may be used, mica or other suitable insulating material being placed between the connectors and the coil. The metal strips, to which the connectors are soldered, should be just long enough to permit of a firm mechanical connection with the coil, by wrapping tape or cord, or both over them.

For particular work, the terminal shown in Fig. 161 is recommended. This may be mounted in the following manner: First place a sheet of Micanite about 14 mils thick between the coil and the terminal, leaving a good margin around the edges to prevent any "jumping" of the current.

FIG. 161.—Terminal Conductor with Water Shield.

The terminals may be firmly held to the coil by the first wrapping of asbestos tape, stout twine being first employed, which is removed as the tape is applied.

The water shield is applied over the asbestos tape or paper before the external insulation is applied. In soldering the terminals on, solder having a melting point of 400° F or more should be used. Never use acid.

Figures 162 to 164 are suggestions for bringing out flexible terminals. Figure 165 shows the method of fastening the inside and outside terminals by means of cotton or asbestos tape. In each case the end of the wire is passed through the loop which loop is drawn together by a sharp jerk on the protruding end of the tape.

FIG. 162.—Method of bringing out Terminal Wires.

Terminals for small electromagnets, such as are used on telephone switchboard apparatus, consist of pins, clips, etc.

A good method of connecting the inside terminals of a pair of coils on a horseshoe electromagnet is to cut a piece of shellacked cotton

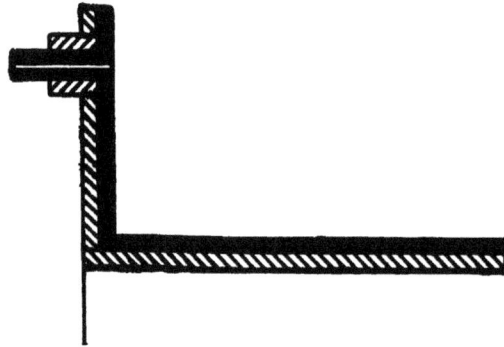

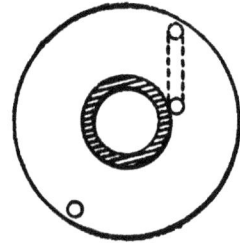

FIG. 163. — Method of bringing out Terminal Wires.

FIG. 164. — Method of bringing out Terminal Wires.

tubing the proper length, and then make a small incision midway between the two ends.   The inside

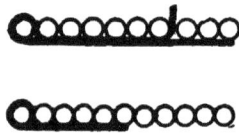

FIG. 165.—Methods of tying Inner and Outer Terminal Wires.

terminal wires are then passed in at opposite ends of the sleeve ; brought out through the incision ; soldered, and tucked back inside of the sleeve. A touch with a brush dipped in shellac seals the incision.   Rubber tubing should never be used unless the wires are tinned, owing to the corrosive effect of the sulphur in the rubber on the copper wire.

CHAPTER XV

INSULATION OF COILS

98. General Insulation

The complete insulation of electromagnetic windings consists of (*a*) the insulation on the wire; (*b*) the internal or extra insulation placed between the layers of wire in the winding, and impregnating compounds for improving the insulation on the wire, and (*c*) the external or insulation placed about the outside of the winding to insulate it from the core, frame, etc. The insulation on the wire is treated in Chap. XVII.

99. Internal Insulation

In all electromagnetic windings there are electrical stresses between adjacent layers and turns. This pressure varies directly with the total voltage across the terminals of the winding, and the length of the winding, and inversely with the number of layers. For this reason it is desirable

Fig. 166. — Sectional Winding.

to keep the length of the winding as small as possible.

Where it is necessary to use a long winding, it is

Fig. 167. — Insulation between Layers.

customary to form the winding proper of several short windings connected in series, and insulated from one another at the ends, as in Fig. 166. For particularly heavy duty, it is also customary to place paper, mica, or insulating linen between the layers of the winding. The paper or other insulating material should project a short distance from each end of the winding, as in Fig. 167. This will prevent any "jumping" around the insulating medium, from layer to layer.

Silk and cotton covered wire coils may be impregnated with varnishes and other compounds specially prepared for the purpose, but enameled wire is better treated by the "dry" process, *i.e.* insulated with oiled linen or mica, according to the temperature at which it is to be operated.

There are three methods of treating the insulated wire to increase the dielectric strength and to make them moisture-proof; the former two, described below, being practically similar.

In one of these methods the wound coil is dipped in an insulating compound until it is saturated as much as possible by the compound or varnish. In the other similar method, the wire is passed through a bath of insulating substance, in liquid form, as it is wound in the coil, or else the layers are painted with the insulating substance, one by one, as they are wound into the coil.

In the third method, the coils are placed in a vacuum chamber and the air pumped out. At the same time the moisture is expelled. When a sufficient degree of vacuum is obtained, a melted insulating compound is allowed to flow into the vacuum chamber, at the pressure of the atmosphere. As no air is allowed to return to the chamber, the insulating compound fills all the interstices between the turns.

In a simple vacuum process, the coil is placed directly in the impregnating compound, and the air is then pumped out. The latter method is not considered so desirable as the former, however, as the moisture is not so readily expelled in the latter method.

Insulating varnish may also be used in the vacuum drying and substituting process, the coil being thoroughly baked afterwards.

In coils designed for use on alternating currents, it is customary to place mica, oiled linen, or paper between the layers, and treat it with some insulating and impregnating compound besides.

The material for the internal insulation of the coil should be oil and moisture proof, chemically inert, and a good conductor of heat. Moreover, it should be mechanically strong, so as to cement the coil in one solid mass, so that the wire cannot vibrate and thus injure its insulation. It should also be unaffected by the heat at the ordinary operating temperature.

Solid and paraffin compounds tend to soften under the influence of heat. Paraffin compounds, while non-hygroscopic, are poor mechanically, and are soluble in mineral oils. For these reasons, oil varnishes are the more preferable. A good varnish should not contain volatile thinners, as these thinners are driven off when the coil is baked, leaving the inside of the coil more or less porous.

The internal insulation of asbestos-covered * wire windings may be affected by dipping the wound coil, while hot, in fairly thick Armalac, density 38° B., allowing the coil to drain a few minutes after each dipping. The coil is preferably heated by suspending it for a short time in an oven, the temperature being raised to about 200° F.

* D. & W. Fuse Co.

100. EXTERNAL INSULATION

The external insulation consists, in the ordinary form of winding on a bobbin, of the insulating sleeve over the core, the insulating washers at the ends, and the wrapping around the outside of the coil. These are usually paper, fiber, hard rubber, oiled linen, mica, etc., depending upon the voltage, and the uses to which they are to be applied.

In many cases the bobbin itself consists of insulating material, in which case only the outer wrapping need be considered.

In any type of bobbin, whether of brass or of fiber or other material of low insulating qualities, the following precautions should be taken to insulate the bobbin for a high voltage. Around the tube place several wraps of oiled linen or similar material, the number of wraps to be at least twice those necessary to resist the voltage, as specified in the table on p. 325. The reason why twice the thickness should be used will be explained presently.

Lay out 3.1416 times the diameter of the tube for the first wrap, and divide this into such a number of parts that the width of each fringe shall be at least $\frac{1}{2}$ inch (this may be less when smaller diameters require it).

For the next wrap, consider the diameter of the tube plus the first wrap as the new diameter, and proceed as before, this time having the same number of segments as in the first layer. This may be carried out as far as desired.

If these directions are carefully followed, the fringed ends will lap, as in Fig. 168. When the linen is in place in the bobbin, the fringed ends will, of course, rest radially against the washers.

A sufficient number of oiled linen or mica washers should now be placed over the fringed ends to prevent

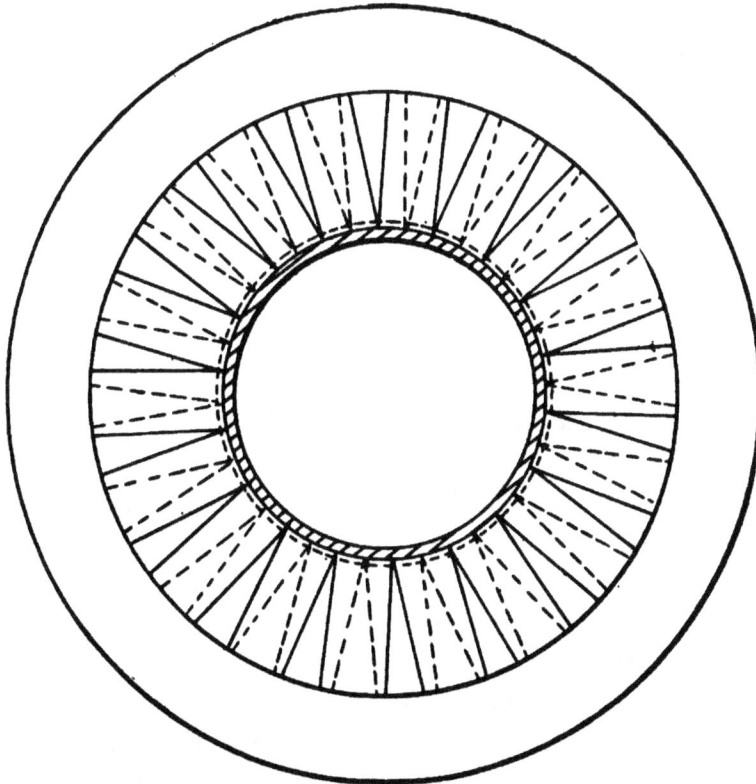

Fig. 168. — Method of mounting Fringed Insulation.

any jumping to the metal washers, and two pressboard washers, one at each end, should be put on to take the wear of the winding, care being taken to have the slits or cuts in the linen washers at least 90° apart, so that there can be no leakage at these points.   It is better to assemble the linen washers before the brass washers, as then the linen will not have to cut.   The linen washers should be placed over the wrapping of linen on the tube, with the fringe between the metal washer and the linen washer. The bobbin will then appear, as in Fig. 169. The corners *a–a* may be painted with varnish, and the whole thoroughly baked.

FIG. 169. — Insulation of Bobbins.

This makes a highly insulated bobbin. By this method, brass bobbins may be used and, therefore, much space saved. The bobbin may also be insulated with micanite, as in Fig. 170.

After the coil is wound and treated, it may be wrapped with oiled linen or canvas in sheet or tape form.

FIG. 170. — Insulation of Bobbins.

Over this may be placed cord or pressboard, for protection. The whole should then be dipped in insulating varnish and thoroughly baked, this operation being repeated, for good results.

Instead of first insulating the bobbin and then winding on the wire, the coil is often wound and insulated separately, before mounting it on the core. While this type is employed on solenoids, lifting and plunger electromagnets, in which case it is usually circular in form, it is particularly used for the field magnets of motors and generators. This type is known as a "former" winding, since it is wound on a collapsible form.

For the external insulation of this type of winding, the treated coil may be covered successively with mica cloth, leatheroid, and linen tape, the whole being dipped in varnish which is oil and moisture proof, and then thoroughly baked.

For asbestos-covered wires, the following * is recommended:

Mix thoroughly, while dry, equal parts of red oxide of iron and powdered asbestos, or asbestos meals, as it is sometimes called, then stir in sufficient Armalac or other approved insulating compound to make up a thick putty or paste. This paste can be readily applied to the surface of the coil with a blunt knife, and should be so applied as to leave the coil quite smooth and free from all air spaces. The object of this material is to prevent the possibility of air pockets which interfere with the radiation of heat, and at the same time to improve the insulation of the coil and increase its water resisting properties.

As this paste, or Dobe, is applied, a first wrapping of pure asbestos, double-selvage edge, woven tape, $\frac{3}{4}$ inch × 0.02 inch thick, should be wrapped around the coil. The best results are secured by winding this into the Dobe as the latter is applied, which in a measure partially impregnates the tape and fills up the spaces between the wrappings.

The coil should now be baked for about an hour, with the object of partially drying the Dobe and at the same time driving off any moisture which the tape might contain. The temperature should be raised slowly to about 200° F., when the coil should be taken out and dipped in Armalac or other approved insulating compound, to thoroughly saturate the tape while hot. It should then be baked for a period of about four hours, the temperature being raised by degrees to at least 400° F.

* D. &. W. Fuse Co.

It should then be removed from the baking oven and allowed to cool off until only slightly warm, when it should be coated with some sort of adhesive varnish such as shellac, or preferably, Walpole gum, which has been found very satisfactory, to bring about a close adhesion between this and the paper covering which is next put on.

The paper jacket consists of a single thickness of 0.03-inch asbestos paper, which is applied in a dampened condition in order to facilitate forming it closely about the coil. It is preferably first cut into shapes which will readily lend themselves to this formation, the edges being secured by any suitable water paste, care being taken to employ only as little of this paste as possible. Over this another layer of asbestos tape is wound.

The coil should now be thoroughly dried out and baked for at least two hours, the temperature being raised slowly to not less than 400° F. While the coil is at its maximum temperature, it should be taken from the oven and immersed for 45 minutes in Delta compound, which has previously been heated to at least 350° F. The coil may be suspended by means of asbestos tape, as it will not burn through.

After removing the coil from the compound, wipe off the surplus and bake for three hours at 350° F. This tends to set the compound thoroughly into the wrapping and to drive off the volatile matter which will prevent the water-proofing material from softening. While the coil is still hot, it should be finished by immersing in any suitable baking Japan, preferably one which will remain in a flexible condition after it has hardened. Two or three coats will add much to the life of the coil.

Coils insulated in this manner have been put to the most severe service imaginable without injury, although at times they have been completely immersed in boiling water for long periods.

Coils may also be thoroughly dried and baked by passing a current of electricity through them. In this case an ammeter and rheostat should be included in the circuit.

CHAPTER XVI

MAGNET WIRE

101. MATERIAL

THE ideal electrical conducting material would be one which would offer no resistance whatever to the electromotive force. In practice, however, the ideal is not attainable. It is, therefore, natural to seek the best conducting material as well as the most economical, using that which is most practicable. This condition is met with copper wire.

Since, in magnet windings, it is desirable to obtain the maximum number of turns, with a minimum resistance in a given winding space, it is readily seen that, while aluminum may be used to advantage in transmission lines, from an economical standpoint, its use in magnet windings is prohibitive, owing to its greater *Specific Resistance* or *Resistivity*.

102. SPECIFIC RESISTANCE

The *Specific Resistance* of a material is the resistance in ohm at 0° C. of a wire one centimeter long and one square centimeter in cross-section.

The specific resistance of pure annealed copper is 1.584×10^{-6}, and for hard-drawn copper, 1.619×10^{-6}. The specific resistance of copper varies with its temperature. The ratio of change in resistance to change in temperature is called the *Temperature Coefficient*.

The chemical purity and mechanical treatment of copper have marked effects upon its properties.

210

103. Manufacture

In the manufacture of copper wire, ingots of commercially pure copper are rolled into rods, and the wire is then drawn from these rods. Rolling improves the structure of copper wires. Hence, the smaller wires may be considered of better quality than the larger sizes.

Commercial copper wires are usually round, although they may be obtained with square and rectangular cross-sections. In this book, round wire will be assumed unless otherwise specified. In all cases a wire should be of uniform cross-section throughout its entire length.

104. Stranded Conductor

Where a conductor of large cross-section is required, and especially if it is to be wound upon a small core, it may be stranded; *i.e.* it may consist of enough smaller wires to make the total cross-section equal to the cross-section of the large conductor. This makes a very flexible conductor. The space factor of a stranded conductor is much larger than that for a solid conductor. For equal diameters, cables or strands have a conducting area of from 20 to 25 per cent less than a solid circular conductor.

A table showing the approximate equivalent cross-sections of wires will be found on p. 307.

105. Notation used in Calculations for Bare Wires

In this book the following symbols are used in connection with bare-wire calculations:

W_B = total weight,
W_l = weight per unit length,
W_c = weight per unit volume,
l_w = length of wire,
ρ_l = resistance per unit length,
ρ_w = resistance per unit weight,
ρ_v = resistance per unit volume,
R = total resistance,
d = diameter of wire,
W_c = 8.89 grams per cubic centimeter,
  = 0.321 pound per cubic inch.

106. Weight of Copper Wire

Since the weight per unit volume, for copper, according to Matthiessen's standard, is 8.89 grams per cubic centimeter, or 0.321 pound per cubic inch, the total weight of any solid mass of copper would be

$$W_B = W_c V, \tag{162}$$

wherein V = volume of copper.

Since, in practice, round wire is used,

$$W_B = 0.7854 \, W_c V, \tag{163}$$

neglecting imbedding of the wires, etc., which is discussed in Chap. XVIII.

107. Relations between Weight, Length, and Resistance

Weight may be found from length or resistance.

$$W_B = l_w W_l \quad (164), \qquad = \frac{R}{\rho_w} \quad (165), \qquad W_l = \frac{W_B}{l_w}. \quad (166)$$

Length may be found from weight or resistance.

$$l_w = \frac{W_B}{W_l} \quad (167), \qquad\qquad = \frac{R}{\rho_l}. \qquad (168)$$

Resistance may be found from weight or length.

$$R = \rho_w W_B \quad (169), \qquad\qquad = l_w \rho_l. \qquad (170)$$

$$\rho_l = \frac{R}{l_w} \quad (171), \qquad\qquad \rho_w = \frac{R}{W_B}. \qquad (172)$$

108. The Determination of Copper Constants

Specific gravity of copper	8.89
1 foot	0.3048028 meter
1 meter	3.28083 feet
1 inch	2.54 centimeters
1 pound	453.59256 grams
1 gram	0.0022046 pound
1 square centimeter	0.155 square inches
1 gram	1 cubic centimeter of water at 4° C.

1 gram of copper $= 1 \div 8.89 = 0.112486$ cubic centimeters.

Therefore, a wire 100 centimeters long, containing 0.112486 cubic centimeter of copper, will be 0.00112486 square centimeter or 0.000174353 square inch in cross-section, and this has been found to have a resistance of 0.141729 International ohm at 0° C. Hence, a wire one foot long and 0.000174353 square inch in cross-section has a resistance of 0.0431991 International ohm at 0° C. From this it follows that a wire 0.001 inch (1 mil) in diameter (0.0000000785398 square inch, or 1 circular mil, in cross-section) will have a resistance of 9.58992 International ohms per foot at 0° C., or 10.3541 ohms per foot at 20° C. or 68° F.

Therefore, in English units,

$$P_F = 30,269 \times 10^{-4}\, d^2, \tag{173}$$
$$P_O = 2924 \times 10^2\, d^4, \tag{174}$$
$$F_P = 33,036 \times 10^{-5}\, d^{-2}, \tag{175}$$
$$F_O = 96,585\, d^2, \tag{176}$$
$$O_P = 342 \times 10^{-8}\, d^{-4}, \tag{177}$$
$$O_F = 103,541 \times 10^{-10}\, d^{-2}, \tag{178}$$
$$O_I = 86,284 \times 10^{-11}\, d^{-2}; \tag{179}$$

wherein P_F = pounds per foot,
P_O = pounds per ohm,
F_P = feet per pound,
F_O = feet per ohm,
O_P = ohms per pound,
O_F = ohms per foot,
O_I = ohms per inch,
d = diameter of wire in inch.

In metric units,

$$K_M = 698 \times 10^{-5}\, d^2, \tag{180}$$
$$K_O = 3186 \times 10^{-4}\, d^4, \tag{181}$$
$$M_K = 1433 \times 10^{-1}\, d^{-2}, \tag{182}$$
$$M_O = 4563 \times 10^{-2}\, d^2, \tag{183}$$
$$O_K = 3138 \times 10^{-3}\, d^{-4}, \tag{184}$$
$$O_M = 2192 \times 10^{-5}\, d^{-2}, \tag{185}$$
$$O_{cm} = 2192 \times 10^{-7}\, d^{-2}; \tag{186}$$

wherein K_M = kilograms per meter,
K_O = kilograms per ohm,

$M_K =$ meters per kilogram,

$M_O =$ meters per ohm,

$O_K =$ kilograms per ohm,

$O_M =$ ohms per meter,

$O_{cm} =$ ohms per centimeter,

$d =$ diameter of wire in millimeters.

109. AMERICAN WIRE GAUGE (B. & S.)

This is the standard wire gauge in use in the United States. It is based on the geometrical series in which No. 0000 is 0.46 inch diameter, and No. 36 is 0.005 inch diameter.

Let $n =$ the number representing the size of wire,

$d =$ diameter of the wire in inch.

Then $\log d = \overline{1}.5116973 - 0.0503535\,n,$ (187)

and $n = \dfrac{\overline{1}.5116973 - \log d}{0.0503535}.$ (188)

n may represent half, quarter, or decimal sizes.

If d represent the diameter of the wire in millimeters,

then $\log d = 0.9165312 - 0.0503535\,n,$ (189)

and $n = \dfrac{0.9165312 - \log d}{0.0503535}.$ (190)

The ratio of diameters is 2.0050 for every six sizes, while the cross-sections, and consequently the conductances, vary in the ratio of nearly 2 for every three sizes.

110. WIRE TABLES

Wire tables showing the diameters, sectional areas, and the relations between weight, length, and resistance, for the various gauge numbers, will be found on pp. 305 and 306.

The following "Explanation of Table" refers to the table on p. 305, and is copied from the *Supplement to Transactions of American Institute of Electrical Engineers.**

"The data from which this table has been computed are as follows: Matthiessen's standard resistivity, Matthiessen's temperature coefficients, specific gravity of copper = 8.89.   Resistance in terms of the international ohm.

"Matthiessen's standard 1 metre-gramme of hard drawn copper = 0.1469 B. A. U. @ 0° C.   Ratio of resistivity hard to soft copper 1.0226.

"Matthiessen's standard 1 metre-gramme of soft drawn copper = 0.14365 B. A. U. @ 0° C.   One B. A. U. = 0.9866 international ohms.

" Matthiessen's standard 1 metre-gramme soft drawn copper = 0.141729 international ohm @ 0° C.

"Temperature coefficients of resistance for 20° C., 50° C., and 80° C., 1.07968, 1.20625, and 1.33681 respectively.   1 foot = 0.3048028 metre, 1 pound = 453.59256 grammes.

" Although the entries in the table are carried to the fourth significant digit, the computations have been carried to at least five figures.   The last digit is therefore correct to within half a unit, representing an arithmetical degree of accuracy of at least one part in two thousand.   The diameters of the B. & S. or A. W. G. wires are obtained from the geometrical series in which No. 0000 = 0.4600 inch, and No. 36 = 0.005 inch, the nearest fourth significant digit being retained in the areas and diameters so deduced.

"It is to be observed that while Matthiessen's standard

* October, 1893.

of resistivity may be permanently recognized, the temperature coefficient of its variation which he introduced, and which is here used, may in future undergo slight revision.

F. B. Crocker,
G. A. Hamilton,
W. E. Geyer,
A. E. Kennely, Chairman,

} Committee on 'Units and Standards.'"

The metric wire table on p. 306 was calculated by the author by means of 6-place logarithms, and carefully checked with a slide-rule, the computations being carried to the fourth significant digit. The last digit is therefore correct to within half a unit.

111. Square or Rectangular Wire or Ribbon

Copper wire is sometimes made square, but wires of this class are usually rectangular in cross-section. These latter wires are commonly called *ribbons*, and are rolled from the standard B. & S. round wires.

In order to calculate one dimension, the other must first be assumed.

If we let ab = cross-sectional area of ribbon,

$$\text{then} \qquad a = \frac{0.7854\, d^2}{b}, \qquad (191)$$

$$\text{and} \qquad b = \frac{0.7854\, d^2}{a}, \qquad (192)$$

and the diameter of a round wire, to have the same cross-section as the ribbon, will be

$$d = 1.128\sqrt{ab}. \qquad (193)$$

In some cases the ratio of thickness to width of copper ribbon is given instead of one of the dimensions. Then if A_w represents the cross-sectional area of the conductor, and p and q are the ratios, and a and b represent the dimensions of the strip, $a : b = p : q$, whence

$$a = \frac{bp}{q}, \qquad (194)$$

and

$$b = \frac{qa}{p}. \qquad (195)$$

Now

$$ab = A_w. \qquad (196)$$

Therefore, $A_w = \frac{qa^2}{p}$ (197), $= \frac{b^2p}{q}$.   (198)

Then $a = \sqrt{\frac{A_w p}{q}}$ (199),   and $b = \sqrt{\frac{A_w q}{p}}$.   (200)

112. Resistance Wires

In the following table * are several resistance wires, with their relative resistances as compared with copper.

Material	Resistance per Mil Foot	Temperature Coefficient per Deg. F.	Temperature Coefficient per Deg. C.	Comparative Resistance
Copper . .	10.3541 @ 68° F.	0.00215	0.00388	1
Ferro-nickel	170 @ 75° F.	0.00115	0.00207	17
Manganin .	248.8	0.00001	0.000018	24
Advance . .	294	Practi-	cally nil	28
S. B. . . .	336	0.000032	0.000058	32
Climax . .	525	0.0003	0.00054	50
Nichrome .	570	0.00024	0.000243	55

* From data furnished by Driver-Harris Wire Co.

For tables giving further information, see pp. 317–321.

Under similar conditions, the carrying capacity of two wires of equal diameter, but of different materials, varies inversely as the square root of their specific resistances.

CHAPTER XVII

INSULATED WIRES

113. The Insulation

In electromagnetic windings it is necessary to insulate the turns from one another, and various means are adopted for this purpose. While, in some windings, bare wire is coiled with a strand of silk or cotton, or merely an air space, between adjacent turns, and with paper between adjacent layers, the usual method is to first cover the wire with some insulating material before coiling it into the winding.

This insulation is a very important factor in electromagnetic windings. An ideally perfect insulation for this purpose should be vanishingly thin, and have a high dielectric strength. Mechanically it should be hard, tough, and elastic. The thickness of the insulation should be uniform. It should be non-hygroscopic, chemically inert, and unaffected by high temperatures.

While there is, at the present, no material which fully satisfies all of the above requirements, there are, nevertheless, several materials in use which are well adapted for this purpose.

114. Insulating Materials in Common Use

Cotton is used to a large extent on the larger sizes of wires where the ratio of insulation to copper will not be great. It is well adapted as a spacer, although it is very hygroscopic, and dielectrically and mechanically weak. It is not adapted for temperatures over 100° C.

Silk is extensively used on the finer sizes of wire, owing to its small factor of space consumption, although the cost is much greater than that of cotton.

Asbestos, while having approximately the same characteristics as cotton mechanically, hygroscopically, and as an insulator, has the desirable ability to withstand high temperatures. It is, however, expensive and occupies somewhat more space than the cotton insulation.

Enameled wire is quite rapidly superseding cotton and silk covered wires, owing to its small space factor, high dielectric strength, and its ability to withstand high temperatures. It is non-hygroscopic.

The enamel has a tendency to become brittle, and to crack when large enameled wires are sharply bent. This, however, has been largely overcome by some manufacturers.

Paper is also sometimes used to insulate wires.

115. METHODS OF INSULATING WIRES

Wire is insulated with cotton, silk, and asbestos, by covering the wire with threads of the insulating material. This is done by automatic machinery in a very economical manner. Paper strips are also wrapped around the wire in a similar manner, the paper being held in place by a suitable varnish or paste, which will cause it to adhere tightly to the wire.

Wire is insulated with enamel by passing it through a long vessel containing enamel in a liquid state, then passing it upward at a uniform speed through a space in which the temperature is automatically maintained constant at about 300° F. by means of gas flames controlled by thermostats. The speed of travel of the wire, the length within the heated space and the temperature are so adjusted according to the thickness of enamel that each particle is thoroughly baked when it passes upward, and the wire is dipped into a second enameling vessel, and so on until three coats of enamel have been placed on the wire. The fourth coat, which is an exceedingly thin one, is applied in the same manner and similarly dried, and gives an excellent finish to the product.

This operation is accomplished by machines, each of which usually handles twelve wires simultaneously.

116. Temperature-resisting Qualities of Insulation

The tests* of magnet wire described below were undertaken as a result of considerable trouble from the field coils of motors breaking down, due to exposure to high temperatures. The motors were mostly on furnace cranes handling molten metal, and the temperature of the cases frequently reached 360° to 400° F.

Referring to Fig. 171, curve A shows the results of a test on No. 16 B. & S. double-cotton-covered magnet wire of 0.060 inch outside diameter. The covering at 280° F. showed discoloration; at 370° F. the covering was smoking badly, and at 472° the wire was completely bare. This wire showed the highest insulation resistance at the start, but fell down after passing 370°.

Curve B refers to a test on No. 16 B. & S. asbestos and single-cotton-covered wire. At 340° a slight smoke showed up; at 372° discoloration of cotton took place; at 460° the cotton burned, and at 506° cotton was gone.

* C. H. Barrett, *Electrical World and Engineer*, Dec. 23, 1905.

From 506° up to 720°, which was the limit of the thermometer in use, the asbestos held good and showed an insulation resistance of 480 megohms at close, against 11 megohms at start. The asbestos, however, would not stand very rough handling, though considering the temperature it was in pretty fair shape. The outside diameter over insulation of this wire was 0.068 inch.

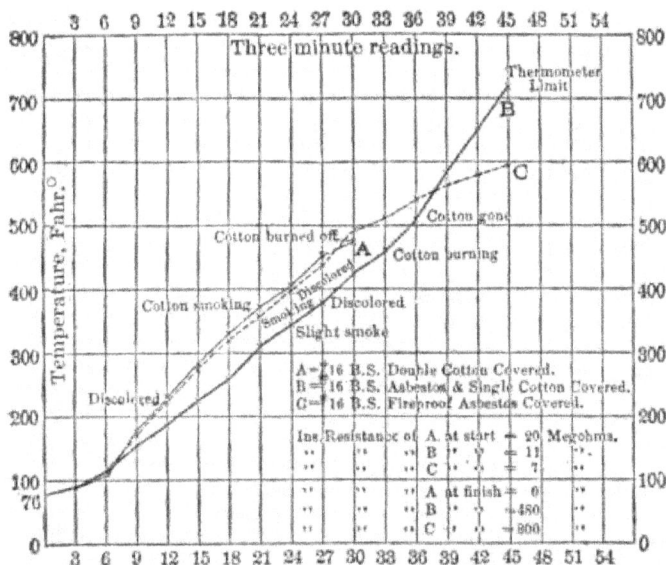

FIG. 171.—Test of Magnet Wire.

In curve C are plotted the data of a test on No. 16 B. & S. fireproof wire which withstood the high temperature excellently except for a slight discoloration, and was in good shape at the finish. Its insulation resistance at start was lower than any, but at the finish reached as high as 800 megohms.

After making the tests it was decided to use the asbestos and cotton-covered wire, and motor troubles were practically ended. Coils have been opened up with every vestige of cotton gone, yet the asbestos kept the insulation almost perfect. The reason the fireproof wire was not used was on account of its affinity for moisture, due, undoubtedly, to the large amount of silicate of soda used in the composition of its covering. In very damp weather the insulation resistance would fall very low on this wire.

The wire in each case was wound on bare sheet-iron spools, then joined in series, and a current of 22 amperes passed through them. A thermometer was placed in each coil and the temperature taken every three minutes. The resistance was taken with a Wheatstone bridge. The thermometers were carefully selected for accurate reading.

Another test,* described below, is of interest.

Investigation has shown that at a temperature of about 147° C., cotton-covered wire will in time char to an extent that will break down its insulation. It was further ascertained that at 199° C. cotton-covered wire will begin to smoke in 20 seconds. At 239° C. it was distinctly discolored in 50 seconds, and complete carbonization had taken place at 245° C. in 2 minutes and 15 seconds.

These temperatures are, of course, excessive, yet they go to show how short a space of time is necessary to ruin the field or armature windings on a railway motor, subjected as they frequently are to enormous overloads. Deltabeston wire tested under identically the same conditions, in fact subjected to identically the same volume of current, is absolutely unaffected.

An interesting comparative test of the properties of

* D. & W. Fuse Co.

the two wires is shown by coupling two pieces together and subjecting them to the same current, resulting in the complete destruction of the cotton insulation while not in the least affecting the Deltabeston wire, which may be further increased in temperature to a dull red heat without its insulation being destroyed.

Thus the only limit to the temperature at which this wire may be run is the oxidation of the copper itself, which will gradually occur if the coil is run continuously at a copper temperature of 250° C.

The fact that the drying process of enameled wire is carried out at a temperature of more than 300° F. is conclusive proof that the enamel will not be injured by any temperature below this value, and some manufacturers claim their enameled wire will not be impaired by a temperature of 500° F.

117. THICKNESS OF INSULATION

The thickness of insulation on an insulated wire is usually referred to as the *increase due to insulation*. As this increase is commonly expressed in mils (the mil being one thousandth of an inch), the increase is usually referred to as *mil-increase*.

For cotton-covered * wires, the mil-increase for the various sizes is usually as follows:

SINGLE-COVERED		DOUBLE-COVERED
Nos. 0000 to 7 . . .	6 mils	12 mils
8 to 19 . . .	5	10
20 to 36 . . .	4	8

For silk-covered * wires:

SINGLE-COVERED		DOUBLE-COVERED
Nos. 16 to 40	2 mils	4 mils

* The Acme Wire Co.

Special silk insulation may be obtained with 1.5 and 3-mil insulation, respectively.

For asbestos-covered wires (Deltabeston):

NOS.	MILS.
0 to 3	18
4 to 7	16
8 to 10	14
11 to 12	12
13 to 20	10

Enamelite wire*:

NOS.	MILS.	NOS.	MILS.
8 to 10 . .	1.6 to 2.0	25 to 27 . .	0.7 to 1.0
11 to 14 . .	1.4 to 1.8	28 to 33 . .	0.5 to 0.7
15 to 19 . .	1.2 to 1.5	34 to 37 . .	0.4 to 0.6
20 to 22 . .	1.0 to 1.2	38 to 40 . .	0.3 to 0.5
23 to 24 . .	0.9 to 1.1		

In any event it is well to caliper the insulated wire with a ratchet-stop micrometer, to ascertain the increase in diameter due to insulation.

In the so-called bare-wire winding, the least distance between the turns of wire, edge to edge, is 3 mils for sizes from No. 34 to No. 40 B. & S. gauge and approximately one half the diameter of the wire for larger sizes. Two sheets of paper, 1 mil thick, are usually inserted between adjacent layers so that the distance between the layers, edge to edge, is 2 mils.

118. NOTATION FOR INSULATED WIRES

In this book, the following notation is used for insulated wires:

W_I = total weight, W_L = weight per unit length,

* The Acme Wire Co.

$W_v =$ weight per unit volume,

$l_w =$ length of wire,

$\rho_l =$ resistance per unit length,

$\rho_i =$ resistance per unit weight,

$\rho_v =$ resistance per unit volume,

$R =$ total resistance,

$i =$ increase in diameter due to insulation,

$d_1 =$ diameter of insulated wire, $= d + i$,   (201)

$d_1^2 =$ sectional area occupied by insulated wire and interstices when coiled into a winding,

$\Sigma =$ sectional area of insulation.

In practice, the value of d_1^2 is equivalent to the square of the diameter of the wire and insulation as measured with a ratchet-stop micrometer, and the charts on pp. 314 to 316 are based on this principle.

119. Ratio of Conductor to Insulation in Insulated Wires

It is evident that

$$\Sigma = 0.7854 \, (d_1^2 - d^2). \quad (202)$$

The percentage of copper in an insulated wire will, therefore, be

$$\text{percentage of copper} = \frac{d^2}{\Sigma + d^2}. \quad (203)$$

For any kind of round insulated conductor the percentage of weight of conductor is

$$W_p = \frac{G_s d^2}{G_s d^2 + g_s \Sigma}, \quad (204)$$

wherein $G_s =$ specific gravity of conductor,

        $g_s =$ specific gravity of insulation.

The values of g_s are approximately as follows: Asbestos 1.6, cotton 1.4, silk 1.0.   Owing to their hygroscopic properties the data obtained from the above materials, when thoroughly dry, are liable to appear rather low.

The weight per unit volume, W_v, for insulated wires may be readily determined by the equation

$$W_v = \frac{\rho_v}{\rho_i}, \tag{205}$$

wherein ρ_v = resistance per unit volume,

and ρ_i = resistance per unit weight.

W_v may be considered as the combined weight and space factor.   (See Fig. 215, p. 309.)

120. INSULATION THICKNESS

When the size of wire and resistance per unit volume are fixed, the required thickness of insulation may be found by the equation

$$i = \sqrt{\frac{c}{\rho_v d^2}} - d, \tag{206}$$

wherein $c = 2192 \times 10^{-7}$ in metric units, and $86{,}284 \times 10^{-11}$ in English units.

Enameled wires are also covered with cotton and silk and are then known as Cottonite and Silkenite.

These wires replace those double-covered with cotton and silk; they have a much higher coefficient of space utilization, and coils wound with Cottonite and Silkenite can be impregnated.   The total thickness of insulation will be the sums of those given on pages 225 and 226 for the various wires.   The turns per square inch approximate the values for S. C. C. and S. S. C. wires.

CHAPTER XVIII

ELECTROMAGNETIC WINDINGS

121. Most Efficient Winding

An electromagnetic winding consists of an assemblage of helices of insulated wire, in a definitely prepared space surrounding the core, the direction of the turns being alternately right and left; that is, the turns do not lie exactly at right angles with the core as they should theoretically.

The most efficient winding is that which has the maximum number of turns of wire for the minimum resistance; consequently that which has the maximum ampere-turns for a minimum voltage.

In an ideal winding, the mass of conducting material would exactly equal the winding space. There would be no space lost due to insulation, which would be infinitesimal, and there would be no interstices between adjacent turns or between adjacent layers.

Even with ideal conditions, however, there could be but two cases where no space would be lost due to the turning back of one layer upon another. In the first case, the winding would consist of but one turn of square or rectangular wire, forming a hollow cylinder, while, in the second case, the winding might consist of an infinite number of turns of square wire whose cross-section should be vanishingly small.

Before departing from the discussion of ideal conditions, which is given to show what a thoroughly practical proposition an electromagnetic winding really is, a comparison of the cross-sections of windings of round and square wires may be appreciated by referring to

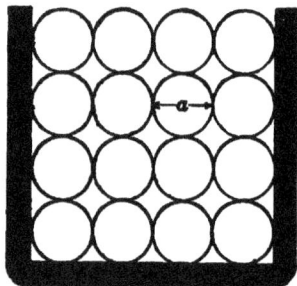

FIG. 172.
Space Utilization of
Round Wire.

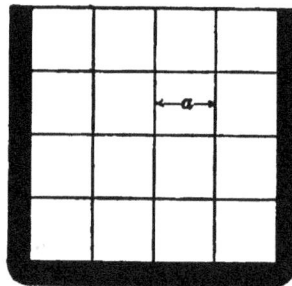

FIG. 173.
Space Utilization of
Square Wire.

Figs. 172 and 173, in which the winding volumes are the same in both cases. For the same number of turns, then, the amount of copper in the winding in Fig. 172 will only contain

$$\frac{\pi}{4} = 0.7854$$

of that in Fig 173, the dimension a being the same in both cases.

In practice, there is no such thing as infinitesimal insulation ; hence, there are interstices between adjacent turns and between adjacent layers.

While wires of square cross-section are sometimes used, in the larger sizes, on the field magnets of motors, etc., and wires or ribbons of rectangular cross-section are also used in certain cases, which will be discussed further on, the magnet wires commonly used are circular in cross-section, and, in this book, the latter form of wire will be assumed unless otherwise stated.

The reason for using a round wire is on account of the tendency of the square wires to lie upon their corners, as well as upon their flat faces, and for the further reason that, as the periphery of a square or rectangle is greater than that of a circle, for equal areas, the extra amount of insulation necessary to cover the wire takes up more of the winding volume for the square or rectangular conductors than for the round wire.

No matter what the form of a winding space may be, there are three dimensions which must always be considered; viz. the average length of all the turns (p_a), the interflange length (L), or the length of the winding, and the depth or thickness (T) of the winding. The volume or cubical contents of any form of winding space may then be expressed

$$V = p_a L T. \tag{207}$$

It may be well to state here that the number of turns in an ideal case are proportional to one half the longitudinal cross-section of the winding, divided by the sectional area occupied by the insulated wire, or

$$N = \frac{TL}{d_1^2}. \tag{208}$$

The turns per unit longitudinal cross-sectional area of winding are

$$N_a = \frac{1}{d_1^2}. \tag{209}$$

Hence, $$N = TLN_a. \tag{210}$$

The resistance may be expressed,

$$R = p_a p_l N, \tag{211}$$

wherein p_a is the average length of all the turns, and p_l is the resistance per unit length of wire.

122. IMBEDDING OF LAYERS

In the round-wire winding, the layers have a tendency to imbed. At the point where the turns of adjacent layers cross one another they appear as in Fig. 172. Diametrically opposite this point there is another

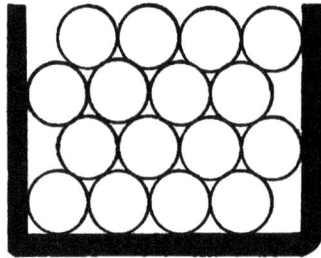

FIG. 174.

Space Utilization of Imbedded
Wires.

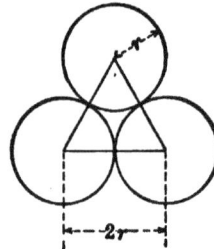

FIG. 175.

Relations of Imbedded
Wires.

crossing point, but at the ends of a diameter at right angles with this one, the turns of the upper layers occupy the space between the layers beneath, as in Figs. 174 and 175.

Theory indicates that there should be a gain of 7.2 per cent in turns on account of this imbedding. However, the insulation is compressed, owing to the vertical tension, which fact causes it to occupy more space latterly than calculated.

As a test * of the imbedding theory, the author had constructed a solid bobbin of steel exactly 2.54 cm. between the faces of the heads, and with a core 1.27 cm. in diameter. This was wound by hand, by an experienced operator, with various sizes of single-silk-covered magnet wire ranging from Nos. 21 to 34 B. & S. gauge. The values for d_1 in Fig. 176 were taken with a ratchet-stop micrometer. This shows the relation (for

* *Electrical World*, Vol. 53, No. 3, 1909, pp 155–157.

eight layers) between thickness of winding T and· calipered diameter of insulated wire. Even with 7.2 per

cent allowed for imbedding, there was found to be an additional "flattening out" of the insulation, due to the vertical compression of the wire, owing to the tension. This averaged approximately 6 per cent.

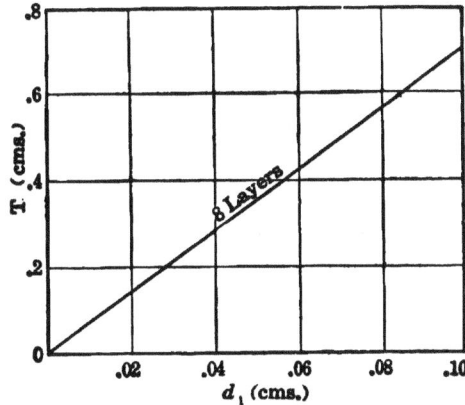

FIG. 176. — Test of an 8-layer Magnet Winding.

For a constant thickness of insulation, it would appear that this effect would vary with different sizes of wire; but since the tension on the wire during the winding process decreases as the diameter of the wire decreases, it remains practically constant for the sizes of wire mentioned above. Examination of the wire when removed from the experimental winding showed that the wire had not been appreciably stretched in winding. This apparent gain of approximately 6 per cent was found to be compensated by a loss of approximately 6 per cent in the turns per unit length.

The formula used for calculating the actual average thickness of the winding per layer is

$$t = \frac{T}{0.933(n-1)+1},\qquad (212)$$

wherein $n =$ the number of layers, and T the thickness or depth of winding.

It will be observed that in this formula an allowance of 7.2 per cent has been made for imbedding.

By transposition,

$$n = 1.072\left(\frac{T}{t} - 1\right) + 1, \qquad (213)$$

and

$$T = t[0.933\,(n-1) + 1]. \qquad (214)$$

123. Loss at Faces of Winding

The loss at the faces or ends of the winding, due to the turning back of one layer upon another, is proportional to the turns per layer. There is a loss of $\dfrac{0+1}{2}$ or one half turn at each end, or one turn per layer.

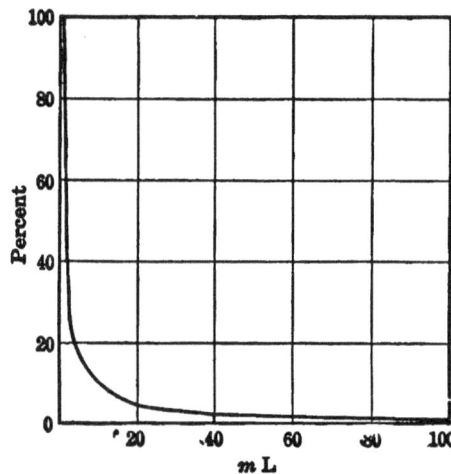

Fig. 177. — Loss of Space by Change of Plane of Winding.

The percentage of loss, due to this effect, is equal to the loss in turns per layer divided by the turns per layer, or

$$\text{per cent loss} = \frac{1}{mL}, \qquad (215)$$

wherein m represents the turns per unit length, and L is the length of the winding. Figure 177 shows that while this loss is not great for small wires, it may be considerable for large wires where L has a small value.

124. Loss Due to Pitch of Turns

Another effect which is very important, and which explains why the insulated wire should be wound evenly in layers, is the loss in magnetizing force, for a given length of wire, when the wire is not at right angles with the core.

All other things being constant, the winding having the highest efficiency will contain the greater number of turns for a given resistance ; but a piece of wire having a given resistance may be so arranged in a corresponding winding space that there will not be one effective turn.

As an extreme case, consider a core wound longitudinally and uniformly with insulated wire. It is apparent that the turns in this case are not effective for magnetizing the core longitudinally, in the ordinary sense.

In an ideal case the conductor would be at right angles to the longitudinal center of the winding, as in Fig. 178, but in all practical windings there is a tendency of the conductors to incline toward the longitudinal center of the winding. This inclination depends upon the diameter of the turn and the diameter of the insulated wire, for layer windings. It is important to always consider the inclination of the average turn, as the inclination is greater for the inner turns, and less for the outer turns, as compared with the diameter of the average turn.

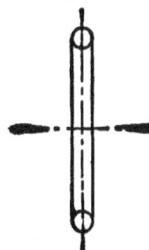

Fig. 178. — Ideal Turn.

In an ideal case the number of turns would be determined by

$$N = \frac{TL}{d_1^2} \quad (208), \quad \text{or} \quad N = TLNa; \quad (210)$$

but, while (208) may hold near enough for many cases in practice, it is important to consider the inclination of the turns, referred to above, when dealing with certain cases.

When the inclination is considered, the number of turns cannot be calculated directly by (208), but the ratio r may be determined by

$$r = \frac{M}{(d_i^2 + M^2)^{\frac{1}{2}}}, \qquad (216)$$

wherein M is the average diameter of all the turns in a round winding, and represents the average perimeter divided by π for any other form of winding space.

Hence, $\qquad M = \frac{p_a}{\pi} \quad (217) \quad = 0.318\, p_a. \qquad (218)$

In "haphazard" or similar windings, the pitch or inclination may be so great that the distance between adjacent turns, which we may designate by d_i, may even exceed M. In this case

$$r = \frac{M}{(d_i^2 + M^2)^{\frac{1}{2}}}. \qquad (216)$$

The number of turns in any winding and with any pitch is

$$N = \frac{rTL}{d_1^2}. \qquad (219)$$

Substituting the value of r from (216) in (219),

$$N = \frac{MLT}{d_1^2(d_i^2 + M^2)^{\frac{1}{2}}}. \qquad (220)$$

When $d_i = M$, the pitch would appear as in Fig. 179. When M is great as compared with d_i, the ratio r will be near unity, but when d_i is greater than M, r has a low value. In Fig. 180 is shown the percentage of turns for various ratios of d_i to M, the size of insulated wire and resistance remaining constant.

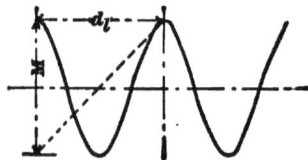

Fig. 179.—Pitch when $d_i = M$.

Figure 180 shows very clearly that an electromagnetic winding should be wound with the turns as close to-

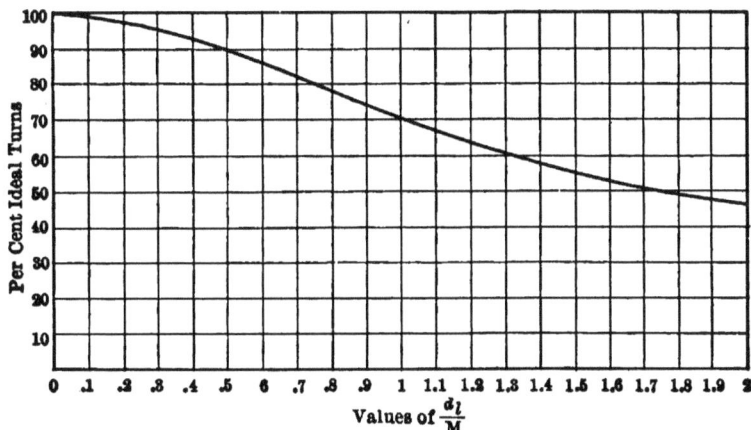

FIG. 180. — Effects Due to Pitch of Winding.

gether, and as near at right angles with the core, as possible.

125. ACTIVITY

It is seen, then, that there are several factors which prevent the mass of conducting material from equaling the entire available winding space. Since a round wire is used in practice, only about 75 per cent of the winding space may be utilized, even with the larger sizes of wire, which represents a loss of approximately 25 per cent. While there is, theoretically, a gain of 7.2 per cent, due to imbedding, this is usually neutralized by deformities in practical windings. Then there is the loss at the ends, due to turning back. This loss may be ignored in fine-wire layer windings, and generally, in windings of considerable length. The inclination of the turns may not be considered in practice, where a uniform, fine-wire layer winding is employed, but this is extremely important in "haphazard" windings.

It is apparent, then, that the thickness of the insulation on the wire is the principal point to be considered in connection with practical round-wire windings, so far as space utilization is concerned.

The *coefficient of space utilization* or *Activity* is the ratio between total cross-section of copper and the total cross-section of winding space. In this it is assumed that the turns are at right angles to the core. Therefore, the practical rule is better expressed as follows :

$$\psi = \frac{0.7854 \, d^2 N}{TL},\qquad (221)$$

wherein ψ is the activity.

In this the total turns are multiplied by the sectional area of the wire, to give the total sectional area of the copper in the winding.

For the ideal winding $\psi = 1$ or 100 per cent. In

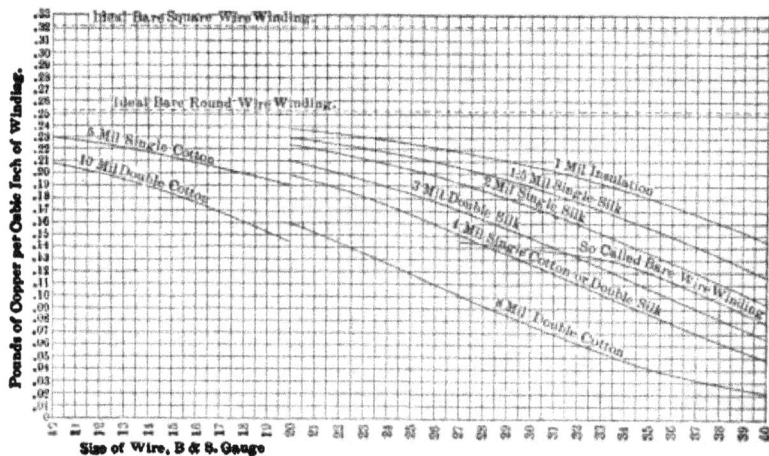

Fig. 181. — Weight of Copper in Insulated Wires.

practice, ψ may be as high as 0.75 with very coarse round-wire field-magnet windings, while in fine-wire coils it may be as low as 0.2, depending upon the thickness of insulation and the regularity of winding.

The space occupied by the insulation on the wires, as well as other interstices, may be appreciated by consulting Fig. 181. It may be noted that a No. 37 B. & S. wire, insulated with silk to a 1.5-mil increase, has twice as much copper per unit winding volume as the same wire insulated with cotton or silk, to a 4-mil increase.

126. AMPERE-TURNS AND ACTIVITY

The ampere-turns in a winding are constant when the size of the wire, length of the average turn (p_a), and voltage across the terminals of the winding are constant, and regardless of the number of turns and, consequently, the activity.

127. WATTS AND ACTIVITY

However, the insulation should be kept as thin as permissible, so as to have as much copper in the winding as possible, as the cost of operating and heating will vary with the actual resistance in the winding, or in any specific case, with the actual weight of copper. Therefore the thinner the insulating material, the less will be the heating, and, consequently, the cost of operating, as heat in a winding is lost energy, and expensive at that.

For this reason, the custom of removing wire from the outside of a winding to reduce the average perimeter, and thus increase the ampere-turns, is poor practice and very inefficient, as the heating is increased many times for only a slight gain in ampere-turns, and the cost of operating increases in exactly the same ratio as the amount of wire in the winding decreases, since watts vary inversely as the resistance, for constant voltage. With constant current, the cost of operating varies directly with the resistance of the winding, but taking off any of the turns would reduce the ampere-turns proportionately.

In a specific case, if 100 volts be applied to a winding consisting of 7620 turns of No. 30 single-cotton-covered wire, with a resistance of 205 ohms, 3710 ampere-turns will be produced at an energy expenditure of 48.7 watts. If now, one half of the turns be removed, leaving 3810 turns, and a resistance of 77 ohms, 4950 ampere-turns will be produced, at an expenditure of 130 watts.

Therefore, to increase the ampere-turns 33 per cent, the cost of operating has been increased 2.67 times; although the cost of the wire has been reduced in the same proportion. Hence, if, say, 20 per cent is saved in the cost of the wire, it will cost 20 per cent more to operate the electromagnet.

128. VOLTS PER TURN

A winding, with internal and external dimensions constant, may be wound with any size of insulated wire, and by varying the voltage across the terminals of the winding, the ampere-turns may be kept constant also.

If bare wire were used, as in the ideal case, or if the ratio of copper to insulation was constant, the volts per turn would also be constant. This may be readily understood when it is remembered that the resistance of the conductor varies inversely as its cross-section, and that the number of turns vary inversely as the cross-section, also. Hence, if a winding contained but one turn of wire, with a resistance of one ohm, and an

e. m. f. of one volt was applied to it, there would result one ampere-turn.   Now if the same space were occupied by two turns, the resistance per turn would be doubled ; *i.e.* the total resistance would be four ohms.   With one volt per turn, the e. m. f. would be two volts ; hence, the current would be one half ampere, and there would be but one ampere-turn, as before.   In practice, the volts per turn vary inversely with ψ.

129. Volts per Layer

What really determines the necessary dielectric strength of the insulation on a wire are the volts per layer, or, to be exact, the e. m. f. between ends of two adjacent layers, as between the points *a–b*, Fig. 182.

Fig. 182. — Showing where the Greatest Difference of Potential Occurs.

Since there are more turns per layer in a fine-wire winding than in a coarse-wire winding, the e. m. f. between adjacent layers will be much greater for the former than for the latter for the same number of turns and volts.   Hence, it is obvious that where fine wires are used, the activity is necessarily less than for coarser wires, although the mechanical properties of the insulation must not be neglected.

The e. m. f. per layer is found by dividing the voltage across the terminals of the winding by the number of layers.   This, however, only gives the average e. m. f. per layer.   What is more important is to find the maximum voltage between any two layers.   This will naturally be at the outer layers.   Hence, to find the maximum e. m. f. between the two outer layers, multi-

ply the e. m. f. between two average layers by the ratio
between the outer and average perimeters, thus,

$$e_m = \frac{2\,Ep_m}{np_a},\qquad(222)$$

wherein e_m is the maximum e. m. f. between the two
outer layers, p_m the mean perimeter of the two outer
layers, n the number of layers, and p_a the average
perimeter of all the layers.

When a winding is to be designed to fill a long winding space, it should be divided into sections so as to
keep the maximum e. m. f. between any two layers as
low as possible. This will be discussed further, in the
proper place.

130. ACTIVITY EQUIVALENT TO CONDUCTIVITY

Thus far the relation of space occupied by the conductor and the insulation covering it have not been

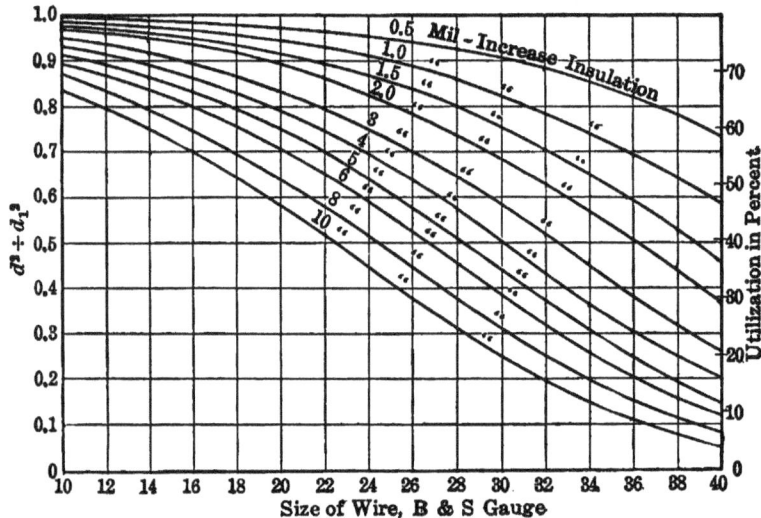

FIG. 183.—Loss of Space by Insulation on Wires.

considered. The activity ratio or practical activity for

round wires is $\dfrac{d^2}{d_1{}^2}$. Figure 183 shows the activity and

the activity ratios for insulated round wires. In this, the other factors, such as imbedding, etc., are not considered.

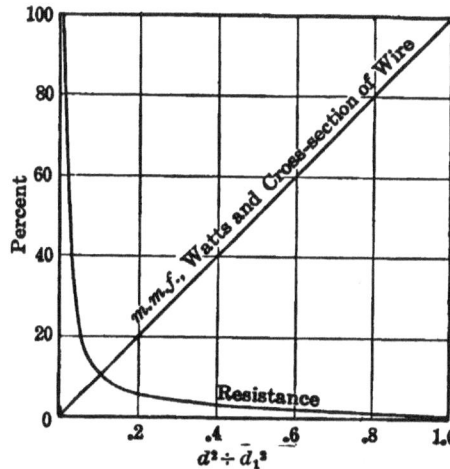

Fig. 184.—Characteristics of Winding of Constant Turns and Length of Wire.

The activity of an electromagnetic winding is equivalent to the conductivity of the conductor itself, where the dimensions of the winding space are limited. This may be appreciated by reference to Figs. 184 and 185. In Fig. 184 the turns and length of wire are constant, and

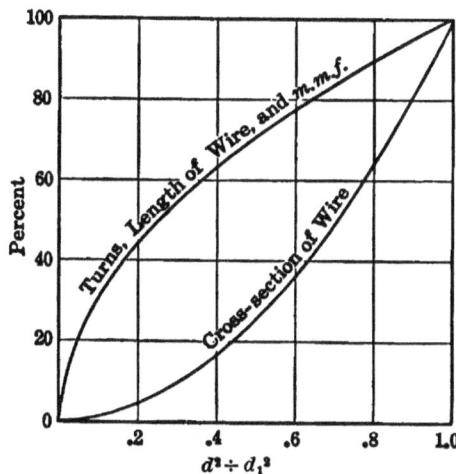

Fig. 185.—Characteristics of Winding of Constant Resistance.

the resistance and size of wire are variable. In this case, if a given winding space be occupied by, say, 5000 turns, with a coefficient of $\dfrac{d^2}{d_1{}^2} = 0.25$, and an exactly similar winding space contains 5000 turns, but with the coefficient of $\dfrac{d^2}{d_1{}^2} = 0.5$, the latter

winding will contain the same number of turns and length

of wire as the former, but will have only one half the resistance, with the cross-section of the wire doubled. Here the size of wire varies directly, and, consequently, the resistance varies inversely as $\frac{d^2}{d_1{}^2}$. Hence, with constant e. m. f., the m. m. f. and watts will vary directly as $\frac{d^2}{d_1{}^2}$.

In Fig. 185 the resistance is constant, and the turns, cross-section, and length of wire are variable. In this case the number of turns and length of wire vary directly as $\sqrt{\frac{d^2}{d_1{}^2}}$, and the cross-section of the wire varies directly as $\left(\frac{d^2}{d_1{}^2}\right)^2$. With constant e. m. f. the m. m. f. will vary directly as $\sqrt{\frac{d^2}{d_1{}^2}}$, and the watts will remain constant.

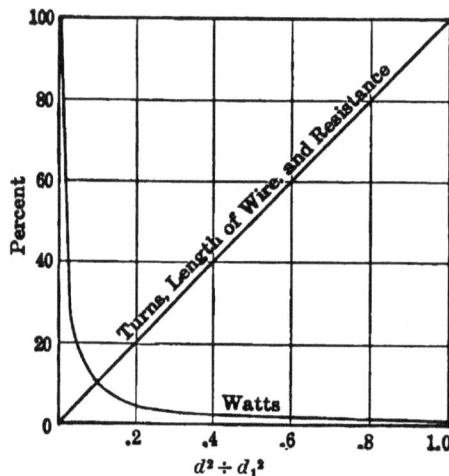

FIG. 186. — Characteristics of Winding of Constant Cross-section of Wire.

In Fig. 186 the cross-section of wire is constant, and the resistance, turns, and length of wire are variable. It is, of course, obvious that the three variables vary directly as $\frac{d^2}{d_1{}^2}$. With constant e. m..f., the m. m. f. will be constant, and the watts will vary inversely as $\frac{d^2}{d_1{}^2}$.

131. Relations between Inner and Outer Dimensions of Winding, and Turns, Ampere-Turns, etc.

The effect of an increased activity is more marked in

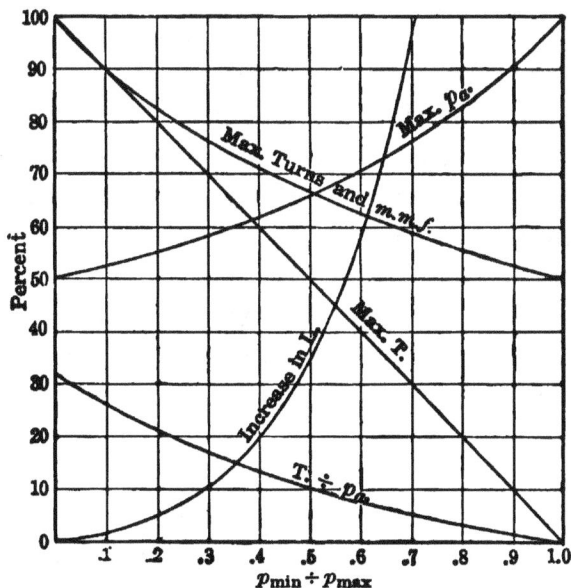

Fig. 187.—Effect upon Characteristics of Windings of Varying the Perimeters.

a winding of small than of large diameter, and varies directly with the length, L, of the winding. Figure 187 shows the various relations, p_{min} and p_{max} being the minimum and maximum perimeters respectively; p_a the average perimeter, and T the thickness or depth of the winding.

132. Importance of High Value for Activity

In order to make the operation of an electromagnetic winding economical, it is readily seen that ψ should have as high a value as possible, since increasing the turns, for a given size of wire, will not change the ampere-turns, but will increase the resistance ; thus reducing the current and, consequently, the cost of operating.

In any case, when designing coils which are to be in use continuously, only the thinnest and best insulation should be used, for the cost of operating will vary in direct proportion to the amount of copper saved by using coarse insulation ; therefore, it pays to use more copper and less current. Moreover, the heating effect decreases as the amount of copper is increased, for the same number of ampere-turns.

When the current is to be on the winding but for a brief period, and when the time between operations is long, the saving in copper is not so important, as the increased cost of the current may not be worth considering.

133. APPROXIMATE RULE FOR RESISTANCE

The resistance of the same kind of insulated wire which will occupy a given winding space varies approximately 50 per cent for consecutive sizes of wire and approximately 100 per cent for every two sizes. This is often convenient for mentally estimating the size of wire to use when the resistance of a similar winding, but with a different size of wire, is known.

134. PRACTICAL METHOD OF CALCULATING AMPERE-TURNS

The following method is convenient for calculating ampere-turns. In this method, use is made of the factor M, which is really the average diameter of a circular winding. In any form of winding, however,

$$M = \frac{p_a}{\pi}. \tag{223}$$

In the American wire gauge (B. & S.) the cross-sectional area of the wires varies nearly in the ratio of 10 for every ten sizes, the real ratio being 10.164 : 1. On this basis Fig. 188 has been plotted, the values for wires from No. 20 to No. 30 being correct; but for wires from No. 10 to No. 20 and between No. 30 and No. 40

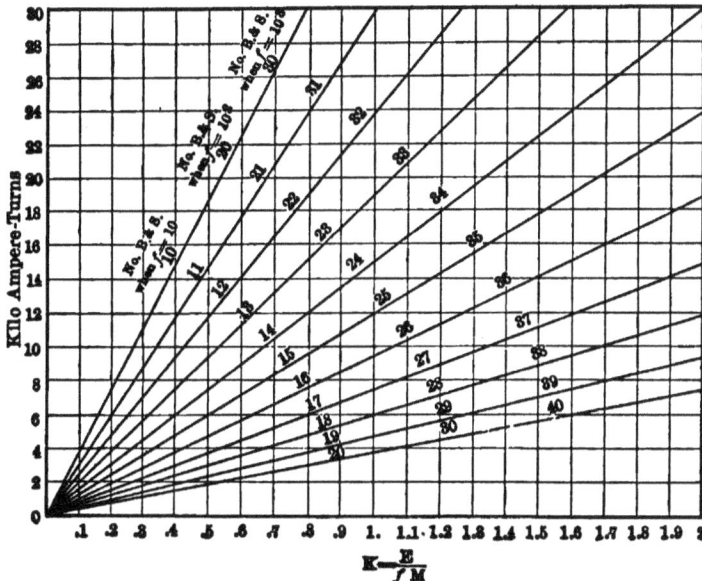

Fig. 188. — Ampere-turn Chart.

the values are correct within 1.64 per cent, which is near enough in practice, owing to the gaps between consecutive sizes of wires.

The ampere-turns may be quickly found by this method in the following manner: First find the ratio Kf by dividing the voltage across the winding by M, or

$$Kf = \frac{E}{M}. \qquad (224)$$

Then by comparing the value of

$$K = \frac{E}{fM} \qquad (225)$$

with the desired ampere-turns, the proper size of wire
(B. & S.) will be found under the value of f, which
value will be either 10, 10^2, or 10^3 for the sizes indicated in Fig. 188. It is well to note here that when
$f = 10$, the values are for wires from No. 10 to No. 20,
and when $f = 10^2$, the values are for wires from No. 20
to No. 30, etc. When the value of

$$K = \frac{E}{fM}$$

exceeds the values of the points of intersection on the
chart, divide this value by 10, and multiply the corresponding value of ampere-turns by 10, or multiply the
value of f by 10, according to whether the size of wire
or ampere-turns is fixed.

The above is deduced from the equation

$$IN = \frac{E}{\rho_l p_a}. \tag{226}$$

135. AMPERE-TURNS PER VOLT

It is often convenient, when designing windings for
different voltages, but for the same type of electromagnet, to estimate the *ampere-turns per volt*. The total
ampere-turns may then be easily calculated from the
total voltage. The chart, Fig. 188, will materially aid
in this operation.

136. RELATION BETWEEN WATTS AND AMPERE-TURNS

It can be shown that the ratio of watts to ampere-
turns is simply the ratio of voltage to turns.

Since $$P = EI \tag{227}, \quad I = \frac{P}{E}. \tag{228}$$

Multiplying both sides of the equation in (228) by N,

$$IN = \frac{PN}{E}, \qquad (229)$$

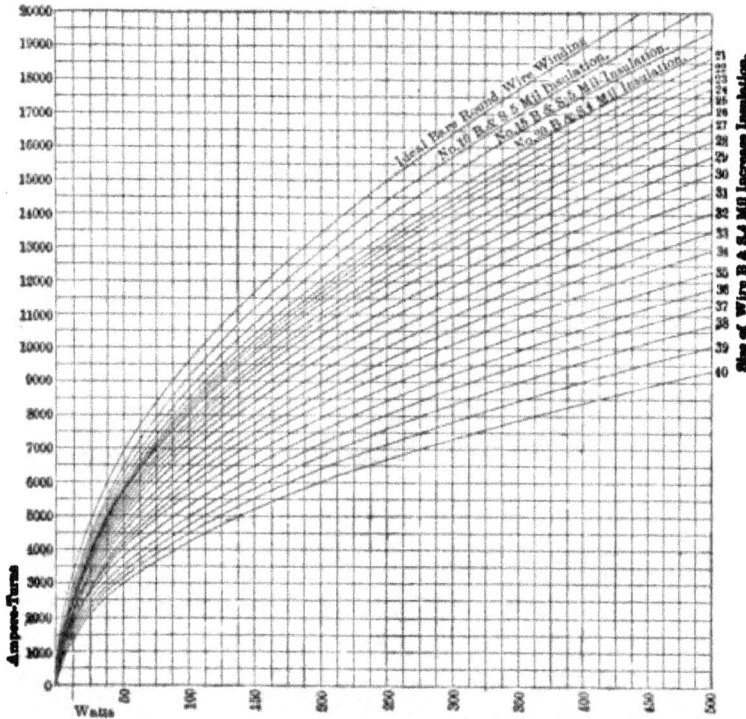

FIG. 189. — Chart showing Ratio between Watts and Ampere-turns.

and since

$$N = \frac{TL}{d_1{}^2} \ (208), \quad IN = \frac{PTL}{Ed_1{}^2} \ (230), \ = \frac{PTLNa}{E}; \quad (231)$$

whence

$$P = \frac{E(IN)}{N} \ (232), \ = \frac{INEd_1{}^2}{TL} \ (233), \ = \frac{INE}{TLNa}. \quad (234)$$

Therefore, the watts may be calculated from the ampere-turns when the other constants are known, and vice versa.

Hence, to calculate watts from ampere-turns, multiply ampere-turns by $\frac{E}{N}$; and to calculate ampere-turns from watts, multiply watts by $\frac{N}{E}$.

Figure 205 shows this relation very nicely. The upper curve represents the theoretical ratio, for a specific case, between watts and ampere-turns for all sizes of wire, no allowance being made for imbedding, etc., and assuming that $d_1^2 = d^2$; *i.e.* $\psi = 0.7854$. It will be noticed that all the other wires shown in curves have 4-mil insulation. Thus to produce 9000 ampere-turns would require an expenditure of approximately 87 watts for any size of round bare wire; 165 watts for a No. 25 wire insulated to a 4-mil increase, and 465 watts for a No. 40 wire insulated to a 4-mil increase.

137. Constant Ratio between Watts and Ampere-turns, Voltage Variable

When it is desired to change the winding of a coil which will produce a required number of ampere-turns at an expenditure of a certain number of watts, with a given voltage, so that it shall produce the same ampere-turns with the same watts on any other voltage, it is necessary, besides using a different size of wire, to change either the average perimeter, the length of the winding, or the thickness of the insulation on the wire, since the ratio of copper to insulation varies with the size of the wire.

The practical method is to change the length, L, of the winding, which will vary inversely with the activity of the winding. Consequently, for fixed average perimeter and voltage, L is proportional to $\frac{\psi IN}{P}$.

138. Length of Wire

The length of wire in a winding is

$$l_w = \frac{V}{d_1^2} \ (235), \text{ or } l_w = VNa, \qquad (236)$$

wherein $V =$ volume of winding space,

and $\quad d_1^2 =$ cross-sectional constant,

$Na =$ turns per unit longitudinal cross-sectional area of winding.

If the length of the wire and the volume of the winding space are known, then the cross-sectional constant may be found by transposition.

$$d_1^2 = \frac{V}{l_w} \ (237), \quad \text{whence } Na = \frac{l_w}{V}. \qquad (238)$$

139. Resistance Calculated from Length of Wire

As the resistance of an electrical conductor of constant cross-section varies directly with its length, it is evident that the resistance of any wire which may be contained in a bobbin or winding volume may be readily calculated by multiplying the length of the wire by the resistance per unit length, ρ_l.

Thus, $\qquad\qquad R = l_w \rho_l. \qquad\qquad (239)$

Values for ρ_l for the various sizes of wires are given in the tables in Chap. XXI.

When using metric units, the *ohms per meter* may, of course, easily be changed to *ohms per centimeter* by simply dividing the former by 10. As the American wire table, in English units, on p. 305 expresses the resistance per unit length as *ohms per foot*, this will have to be divided by 12 to reduce it to *ohms per inch*.

Likewise, the *kilograms per meter* and *pounds per foot* must be reduced to the same units as used in calculating the dimensions of the winding space.

140. RESISTANCE CALCULATED FROM VOLUME

Since the length of the wire $= l_w = \dfrac{V}{d_1{}^2}$ (235), or

$l_w = VNa$ (236), (239) becomes $R = \dfrac{V\rho_l}{d_1{}^2}$ (240), or

$R = V\rho_l Na$ (241).

When $V = 1$, $R = \dfrac{\rho_l}{d_1{}^2}$; hence, it is evident that $\dfrac{\rho_l}{d_1{}^2}$ represents the *resistance per unit volume* to which ρ_v is assigned.

Therefore, $\rho_v = \dfrac{\rho_l}{d_1{}^2}$ (242), or $\rho_v = \rho_l Na$.   (243)

It is then a simple matter to calculate the resistance, when the other constants are known, by multiplying the volume of the winding by the resistance per unit volume :

Thus, $R = \rho_v V$.   (244)

For values of ρ_v see charts, pp. 314–316.

The proper value for ρ_v, to produce the required resistance in a given winding volume, may be determined by rearranging (244), whence $\rho_v = \dfrac{R}{V}$.   (245)

The charts, pp. 314–316, show the *ohms per cubic inch* for various diameters of copper wire, irrespective of the gauge number, with various increases in diameter due to insulation. For convenience, the different sizes of wire of B. & S. gauge are shown in dotted lines, in positions corresponding to their diameters.

As an example of the use of these charts, refer to Fig. 218, and assume that an insulated copper wire is desired which shall have a resistance of 4 ohms per cubic inch when wound on a bobbin.

Tracing vertically upward from 4, it will be found that this result is obtained with a wire 0.018 inch in diameter, with 8-mil insulation, or with a wire 0.0184 inch in diameter with 7-mil insulation, etc., the largest diameter of copper being obtained with 1.5-mil insulation, the diameter of the wire being 0.0208 inch.

Therefore, if the 8-mil insulation be used, a No. 25 B. & S. wire would be used, while with even 3-mil insulation a No. 24 B. & S. wire would suffice, this latter wire being desirable.

Likewise, if the bobbin will contain 1.24 cubic inches of wire, and a resistance of 5000 ohms is required, it is evident that an insulated wire with 4050 ohms per cubic inch would satisfy this condition, and by referring to Fig. 221 it is found that No. 40 B. & S. wire with 1.5-mil silk insulation will meet this requirement.

141. RESISTANCE CALCULATED FROM TURNS

When the number of turns, size of wire, and average perimeter are known,

$$R = \rho_l \, p_a N. \tag{246}$$

The size of insulated wire and the resistance may be determined when the dimensions of the winding space and number of turns are known by first finding the value

$$Na = \frac{N}{TL}. \tag{247}$$

The next smaller size of wire should be selected

from the table, and a new value calculated by the formula

$$V = \frac{p_a N}{Na}.$$ (248)

The resistance will then be

$$R = \rho_v V.$$ (244)

142. Exact Diameter of Wire for Required Ampere-turns

Since $\rho_l = \dfrac{c}{d^2}$ (249) (see page 228 for values of c)

and $IN = \dfrac{E}{\rho_l p_a}$ (226),

$$d = \sqrt{\frac{cINp_a}{E}}.$$ (250)

143. Weight of Bare Wire in a Winding

The weight of copper in a winding may be calculated from the activity by the formula,

$$W_B = VW_c\psi,$$ (251)

where W_c is the weight per unit volume for bare wire, *i.e.* for solid copper.

In metric measure, $W_c = 8.89$ grams per cubic centimeter.

In English measure, $W_c = 0.32$ pound per cubic inch.

The weight may also be found by dividing the resistance by the resistance per unit weight, ρ_w, for bare wires.

Thus,

$$W_B = \frac{R}{\rho_w}.$$ (252)

Also,

$$W_B = l_w W_l.$$ (253)

144. WEIGHT OF INSULATED WIRE IN A WINDING

By substituting the weight factors for the resistance factors, in any formula, the weight of insulated wire in a winding may be obtained.

Thus, $W_I = \dfrac{VW_L}{d_1^2}$ (254), or $W_I = VW_L Na$, (255)

wherein W_L is the weight per unit length.

$W_v = \dfrac{W_L}{d_1^2}$ (256) = weight per unit volume for insulated wires.

Therefore, $W_I = VW_v$. (257)

The weight may also be obtained by dividing the resistance by the resistance per unit weight.

Thus, $W_I = \dfrac{R}{\rho_l}$ (258), or $W_I = \dfrac{RW_v}{\rho_v}$. (259)

Also, $W_I = l_w W_L$. (260)

145. RESISTANCE CALCULATED FROM VOLUME OF INSULATED WIRE

The resistance may be calculated from the weight values in Fig. 181, or from the activity, by comparing the weight of a solid mass of copper having the same volume as the winding, and the actual weight of copper in the insulated wire constituting the winding.

If calculated from weight,

$$R = \rho_w W_B,$$ (261)

wherein ρ_w = ohms per unit of weight for bare wires. (See table, p. 308).

If calculated from activity,

$$R = \psi VW_c \rho_w,$$ (262)

wherein W_c = weight of copper per unit volume. (See Fig. 181.)

146. DIAMETER OF WIRE FOR A GIVEN RESISTANCE

To find the exact diameter of wire to use in a given case, when the increase due to insulation is known, use the formula

$$d = \left[\sqrt{\frac{cV}{R} + \frac{i^2}{4}} \right]^{\frac{1}{2}} - \frac{i}{2}. \tag{263}$$

147. INSULATION FOR A GIVEN RESISTANCE

The increase due to insulation may be determined for a special case, by the formula

$$i = \sqrt{\frac{cV}{Rd^2}} - d. \tag{264}$$

In the above, $c = 2192 \times 10^{-7}$ for metric measure, and $86{,}284 \times 10^{-11}$ in English measure.

CHAPTER XIX

FORMS OF WINDINGS AND SPECIAL TYPES

148. CIRCULAR WINDINGS

THE average perimeter of the winding is

$$p_a = \pi M, \qquad (265)$$

wherein $M =$ average diameter of the winding.

Hence, $V = \pi MLT,$ (266)

wherein $T =$ thickness of the winding,

and $L =$ length of the winding.   (See Fig. 190.)

$$M = \frac{D + D_1}{2} \quad (267),$$

$$T = \frac{D - D_1}{2} \quad (268),$$

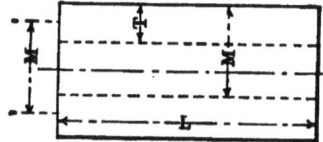

FIG. 190.—Winding Dimensions.

wherein $D =$ outside diameter of the winding,

and $D_1 =$ diameter of core + insulating sleeve, or true inside diameter of the winding.

Substituting the value of V from (266) in (235),

$$l_w = \frac{\pi L(D^2 - D_1^2)}{4\,d_1^2} \quad (269), \quad = \frac{\pi MLT}{d_1^2} \quad (270)$$

$$= \pi MLTNa. \qquad (271)$$

Then, $R = 0.7854\,\rho_v L(D^2 - D_1^2)$ (272)

$$= \rho_v \pi MLT. \qquad (273)$$

From (273) it follows that

$$\rho_v = \frac{R}{\pi MLT} \quad (274), \quad = \frac{1.273\,R}{L(D^2 - D_1^2)}. \qquad (275)$$

Referring to the charts, pp. 314–316, select the next smaller size of wire or next greater value for ρ_v (ohms per cubic inch), and calculate the actual diameter to wind to by the formula

$$D = \sqrt{\frac{1.273\,R}{\rho_v L} + D_1{}^2}. \qquad (276)$$

To find the internal diameter of the winding, under similar conditions, when the outside diameter D is fixed, use formula derived from (276),

$$D_1 = \sqrt{D^2 - \frac{1.273\,R}{\rho_v L}}. \qquad (277)$$

The thickness or depth of the winding for a given volume will be,

$$T = \sqrt{\frac{V}{\pi L} + \frac{D_1{}^2}{4}} - \frac{D_1}{2}. \qquad (278)$$

Substituting $\dfrac{R}{\rho_v}$ for V in (278),

$$T = \sqrt{\frac{R}{\rho_v \pi L} + \frac{D_1{}^2}{4}} - \frac{D_1}{2}. \qquad (279)$$

By this method, the depth of the winding may be calculated for a standard size of wire, when the other factors are given.

The volume of a winding may be quickly approximated by use of the chart (Fig. 191), which will give the value of πMT, and then multiplying by L.

Referring to Fig. 191 the winding volume (in cubic inches) per inch of length of winding is found by following the curved line, which starts from the value of D_1, the inside diameter, to where it intersects the horizontal line corresponding to the value of D, the outside diameter, and then tracing vertically downward.

As an example, the outside of a winding is 2 inches and the diameter of the insulated core, D, is 0.9 inch. Following the curve which starts at 0.9 it will be found that it intersects the horizontal line corresponding to 2 at the

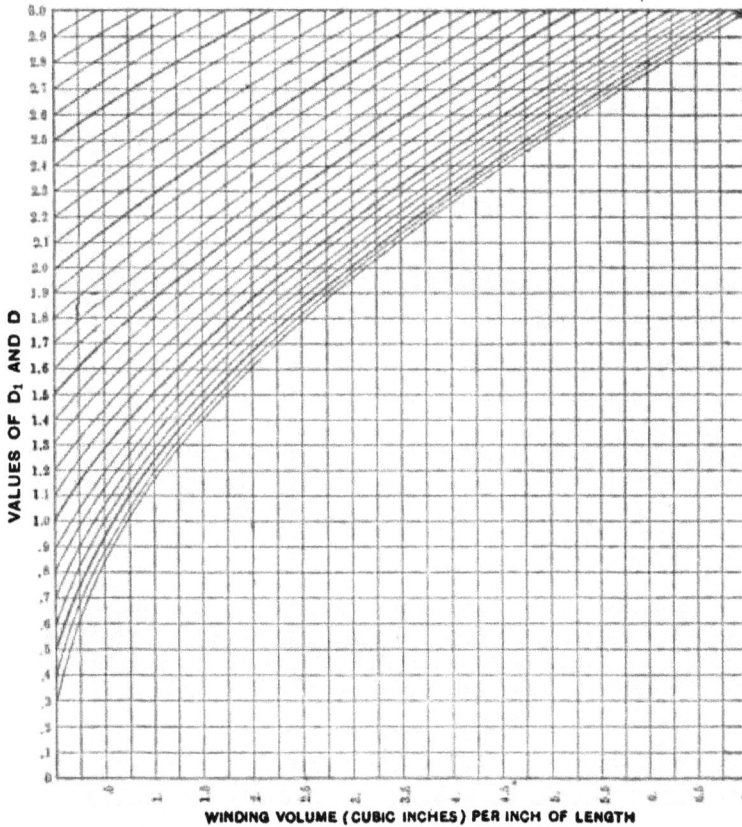

FIG. 191.—Chart for Determining Winding Volume.

vertical line corresponding to 2.5 cubic inches per inch of length of winding. If the length L be 3 inches, the volume of the winding will then be $2.5 \times 3 = 7.5$ cubic inches.

The superficial area (not including ends) is

$$S_r = \pi D L. \qquad (280)$$

The area of each end is $A = p_a T$ (281), $= \pi M T$. (282)

149. Windings on Square or Rectangular Cores

In the calculation of windings with cores of square or rectangular cross-section, the form of the winding is,

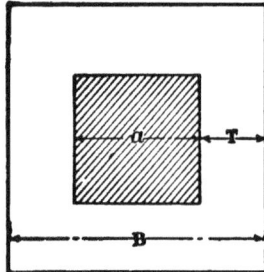

Fig. 192. — Imaginary Square-core Winding.

generally, assumed to be as shown in Fig. 192, and its cross-sectional area is calculated accordingly. That this method is very impracticable will be appreciated by any engineer who may have calculated a winding by it, and compared the actual characteristics of the wound coil with his theoretical deductions.

Most text-books express the volume of such a square-core winding by the formula $V = L(B^2 - a^2)$, wherein a and B are as in Fig. 192, and L is the length of the coil. It would be an extremely difficult operation to cause all the turns to form perfect squares; this would be possible only with exceedingly fine wires.

All practical square-core windings, such as are used on field magnets, etc., appear as in Fig. 193. This is perfectly natural, since the tension on the wires, during the winding process, tends to press the layers closely together at the corners; hence, the thickness, or depth T, of the winding will be uniform all around the core, excepting at the sides, where it has a tendency to bulge outward, owing to the re-

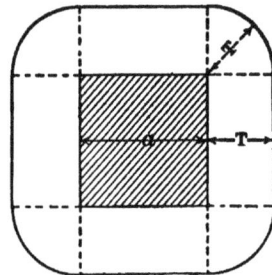

Fig. 193. — Practical Square-core Winding.

siliency of the wire, and the lack of pressure on the flat faces. This latter effect, however, need not be considered in practice.

For a practical square-core winding, as in Fig. 193,

$$p_a = 4\,(a + 0.7854\,T). \qquad (283)$$

Assuming $T = \dfrac{a}{2}$, approximately 7 per cent less wire will be required in the practical type (Fig. 193) than in the imaginary winding (Fig. 192) to accomplish the same results with the same size of core, wire, and number of turns. This is due to the decrease in the average perimeter of the winding, in the practical type.

As a matter of fact, the cores of this class of electromagnet usually have their sharp edges rounded off. We may, then, consider Fig. 194 to be a fairly accurate representa-

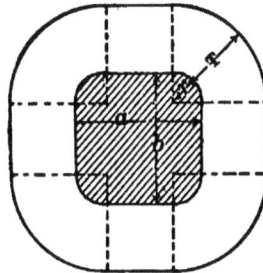

FIG. 194.—Winding on Core between Square and Round.

tion of the cross-section of windings of this nature. An inspection of Fig. 194 will show the average perimeter of the winding to be

$$p_a = \pi(T + 2\,r) + 2\,(a - 2\,r) + 2\,(b - 2\,r), \qquad (284)$$

or $\quad p_a = 2\,(a + b) + \pi T - 1.717\,r, \qquad (285)$

for windings with cores of square or rectangular cross-section.

Since, for square-core windings, $a = b$, (285) may be written $\quad p_a = 4\,a + \pi T - 1.717\,r, \qquad (286)$

or $\qquad\qquad p_a = 4\,(a + 0.7854\,T - 0.423\,r). \qquad (287)$

(287) also holds for circular windings, as in Fig. 195, when $a = 2\,r$.

Referring to Fig. 194,

$$T = \frac{B - a}{2}\;(288), \; = \frac{B_1 - b}{2}. \qquad (289)$$

Then, $B = 2T + a$ (290), and $B_1 = 2T + b$, (291)

wherein B and B_1 = outside dimensions of winding.

The volume of the winding will then be

$$V = p_a TL \text{ (292)} = TL(4a + \pi T - 1.717\,r) \text{ (293)}$$

for windings on square cores, or

$$V = TL[2(a + b) + \pi T - 1.717\,r] \quad \text{(294)}$$

for windings on rectangular cores.

Substituting the value of T from (288) in (294)

$$V = L\left(\frac{B-a}{2}\right)\left[2(a+b) + \pi\left(\frac{B-a}{2}\right) - 1.717\,r\right], \text{(295)}$$

or $V = L(B-a)[(a+b) + 0.7854(B-a) - 0.859\,r]$,

$$\text{(296)}$$

for either square or rectangular windings.

The following formulæ are here repeated for convenience. By substituting the values of p_a, T, V, etc., as given above, the resistance, turns, weight, etc., may be readily calculated.

$$l_w = VN_a \text{ (236)}, R = l_w \rho_l \text{ (239)}, = \rho_v V \text{ (244)},$$

$$N = TLN_a \text{ (210)}, IN = \frac{E}{\rho_l p_a}\text{(226)}, \rho_v = \frac{R}{V}. \quad \text{(245)}$$

For square-core windings, when the value of r is small enough to be neglected,

$$T = \sqrt{\frac{V}{\pi L} + 0.406\,a^2} - 0.637\,a. \quad \text{(297)}$$

Since $$V = \frac{R}{\rho_v}, \quad \text{(298)}$$

$$T = \sqrt{\frac{R}{\rho_v \pi L} + 0.406\,a^2} - 0.637\,a. \quad \text{(299)}$$

By formula (299) the thickness of the winding, for a

standard size of insulated wire, may be calculated from the resistance, when the other constants are known.

The superficial area (not including ends) is

$$S_r = 2L(0.7854\,B + 0.215\,a + b - 2.43\,r). \quad (300)$$

The area of each end is $A = p_a T.$ (281)

150. Windings on Cores whose Cross-sections are between Round and Square

In most cases the space for the winding is of such a nature that its periphery may be either a circle or a square. The dimension B (see Fig. 192) will, therefore, be the limiting dimension for either form of winding; consequently, it is important to determine what form of core and winding will give the best results for any given case.

In this particular case (see Figs. 193 and 195) the dimension a repre-

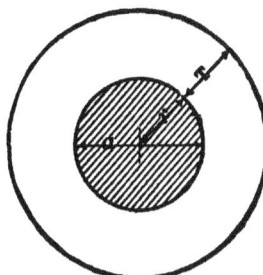

Fig. 195. — Round-core Winding.

sents either the diameter of a round core, or one side of a square core.

For equal areas, the perimeters of the circle and the square are to each other as 1 : 1.128. Hence, if it were possible to construct a winding whose thickness, or depth, would be zero, the economy of the round-core magnet would be 12.8 per cent greater than that for the square-core magnet. However, the thickness of the winding changes the ratio of average perimeters; thus, in two windings, one with a core 1 cm. square, and the other with a round core 1 sq. cm. in cross-section, the thickness of each winding being 10 cm., the economy of the round-core winding would be only 1.3 per cent greater than that with the square core. Therefore, for equal areas, when $T = 0$, the round-core winding has the maximum economy; but, when $T = \infty$, the economies of the square-core and the round-core windings would be the same.

The dimension a, however, will be 12.8 per cent greater for the round core than for the square core.

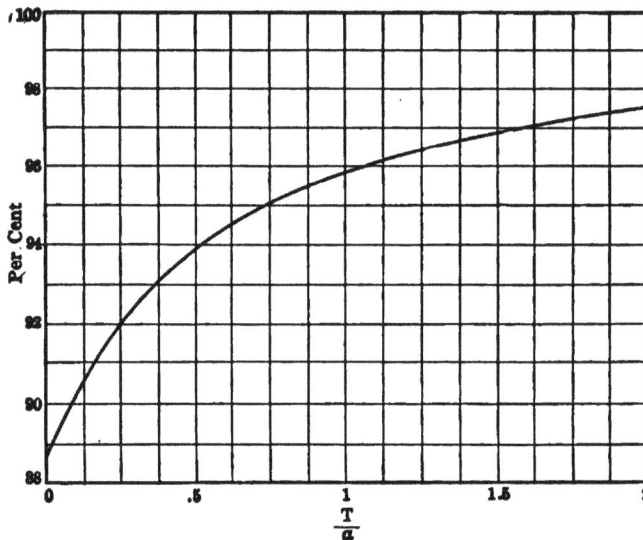

Fig. 196. — Relations between Outside Dimension B of Square-core Electromagnets, and Outside Diameter of Round-core Electromagnets.

This will greatly increase the outside dimensions of the finished coil, where a round coil is used; providing, of course, that the thickness of the winding is not great as compared with a. Figure 196 shows this relation; the outside dimension B being compared, for equal core areas with the outside diameter of the round-core winding. In this particular case, the value of a for the square-core winding is taken in order to have the winding thickness the same for both the square-core and round-core windings.

On the other hand, if we make the dimension a constant for both forms of cores, and the thickness of the winding equal to $10\,a$, the average perimeter of the square-core winding would be 2.5 per cent greater than that for the round-core winding; but, the cross-sectional area of the square-core would be 27 per cent greater than that for the round core.

In any case the flux-density per square centimeter is expressed by the formula

$$\mathscr{B} = \frac{4\,\pi A_c \mu IN}{10\,l_c}, \tag{301}$$

wherein A_c is the cross-sectional area of the core in square centimeters, μ the permeability, I the current in amperes, N the number of turns of wire in the winding, and l_c the length of the magnetic circuit. The ampere-turns are expressed by equation (226),

$$IN = \frac{E}{\rho_l p_a}.$$

Substituting the value of IN from (226) in (301),

$$\mathscr{B} = \frac{4\,\pi A_c \mu E}{10\,l_c \rho_l p_a}. \tag{302}$$

Assuming the values of μ, E, l_c, and ρ_l to be constant, the value of $\mathscr{B}$ will vary directly with the ratio $\dfrac{A_c}{p_a}$.

While the practical round-core electromagnet has the greater economy, magnets with square cores are, nevertheless, extensively used. When the dimension a of the core and the outside dimension B of a square-core electromagnet are fixed, its economy may be considerably increased by rounding the corners of the core, as in Fig. 194. It will be seen that, by increasing the value of r from 0 to $0.5\,a$, the square core, by gradual

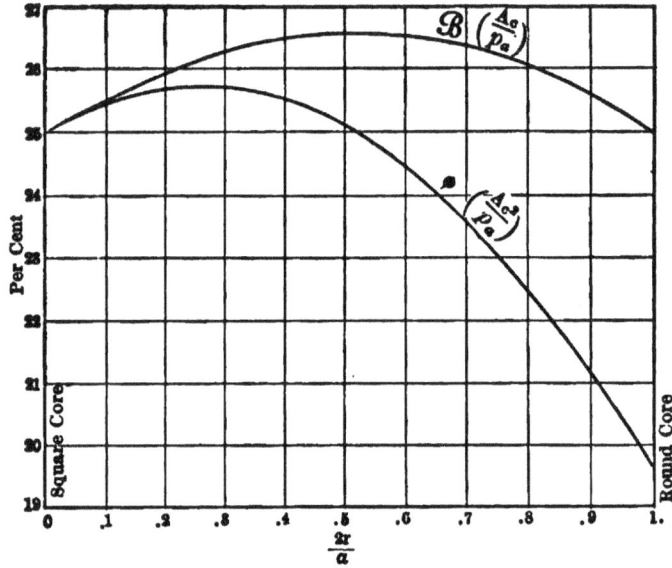

FIG. 197. — Ratios between Round-core and Square-core Electromagnets when $\frac{T}{a} = 0$.

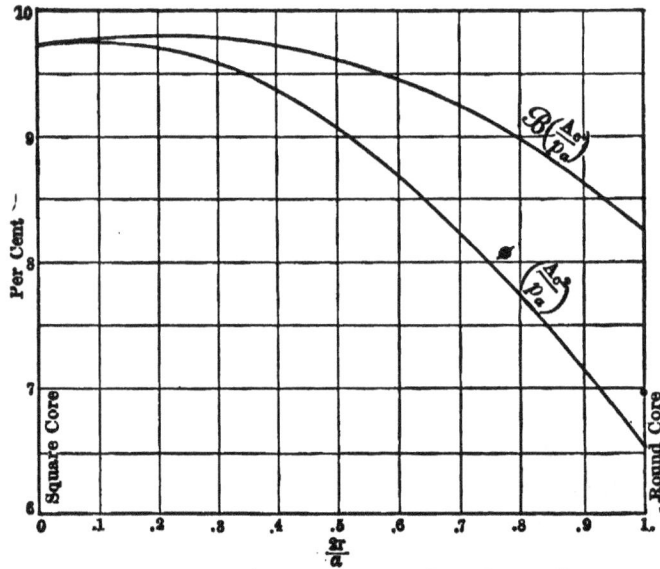

FIG. 198. — Ratios between Square-core and Round-core Electromagnets when $\frac{T}{a} = 2$.

transition, becomes perfectly round; a remaining constant. It is, therefore, obvious that, for various ratios of T and a, the core and winding, for maximum economy, will fall somewhere between the square core and the round core.

In Fig. 197 is shown the ratios of $\dfrac{A_c}{p_a}$ for flux density, and $\dfrac{A_c{}^2}{p_a}$ for the total flux, when $\dfrac{T}{a} = 0$. Figure 198 shows the relation when $\dfrac{T}{a} = 2$. The maximum values for $\mathcal{B}$ and ϕ are shown in Fig. 199. The values of μ, E, l_c and ρ_l, as before stated, are assumed to be constant.

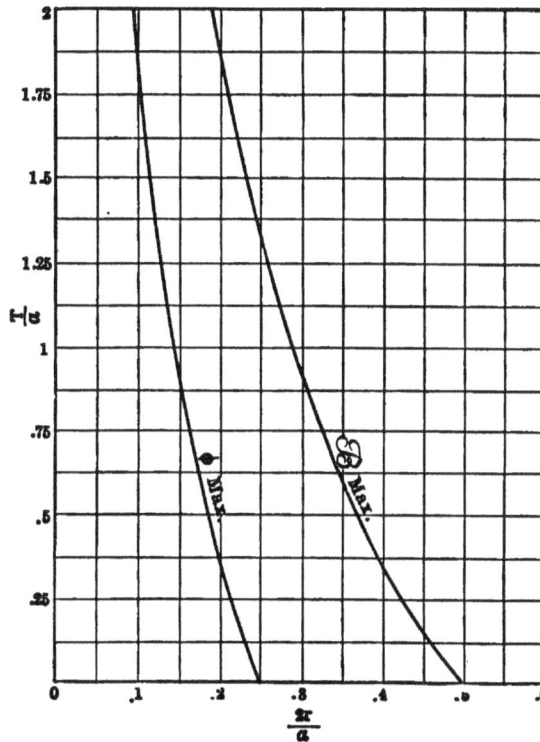

Fig. 199. — Maximum Values for Flux Densities and Total Flux, and Ratios between Core Area and Average Perimeters.

The dimension a is also assumed to remain constant, while the values of r and T are variable. This shows the ratios of the various average perimeters and core areas, taking the cross-sectional area of the square core as unity for the core areas, and the average perimeter of the square-core winding as unity for the average perimeters.

When $\dfrac{T}{a} = 0$, $a = B$.　Hence, the maximum core area
will be obtained with the square core.　The round-core
winding, however, has the minimum average perimeter,

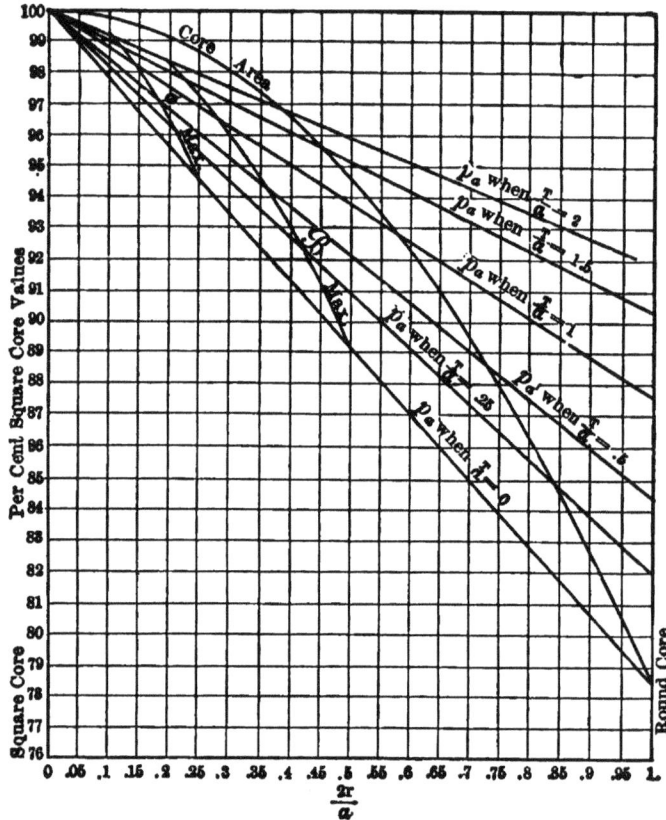

FIG. 200.—Maximum Flux Density and Total Flux, for Various Values of
$\dfrac{2r}{a}$ and $\dfrac{T}{a}$.

and, as the ratio between core area and average perimeter is the same, the value of $\dfrac{2r}{a}$, which will produce
the maximum flux density, will be found to be 0.5.

The condition which shall produce the maximum flux is met when $\dfrac{2\,r}{a} = 0.25$.

The proper value of $\dfrac{2\,r}{a}$, to produce maximum results, for practical windings, will be found at the points where the flux density or the total flux curves intersect with the average perimeter curves. This relation is shown more clearly in Fig. 200.

The usual insulation between the core and the winding has not been considered. However, the average perimeter of the winding may be compared with the total cross-sectional area of the core, plus the insulating medium, and corrections made for the difference between the actual and assumed core areas. In any case, it is important that the average perimeter shall be calculated for the actual winding only.

151. OTHER FORMS OF WINDINGS

For any form of winding volume, simply find the average perimeter p_a, the thickness of the winding T, and the length L. From these dimensions all necessary calculations may be made by using the formulæ in Chapter XVIII.

152. FIXED RESISTANCE AND TURNS

Some specifications call for a certain resistance, or else the weight of the insulated wire is specified. This is often done to keep a check on the manufacturer, to see that the full amount and proper grade of insulated wire is being supplied. Also, in certain apparatus, and particularly that used in telephone switchboards, the magnets are so adjusted that they will only operate above certain current strengths, and as these are connected partly in series and partly in multiple in the same circuit, with a fixed voltage operating the entire combination, it is extremely important that the resistance and turns should be as near a fixed standard as possible for each electromagnet.

153. Tension

In the practical winding of an electromagnet, the tension is a most important factor, for if the wire be wound tightly in a bobbin with fixed winding space, and the same kind of wire be wound loosely upon another identical bobbin, less turns will be obtained in the latter than in the former; consequently there will be less wire in the winding. Hence, with voltage constant in both cases, there will be less current consumption in the former than in the latter case for the same ampere-turns.

It is, therefore, important to use a device which shall keep the tension on the wire constant. By such a means the tension may be so adjusted that all of the windings will be almost absolutely identical both as to turns and resistance.

When winding fine wires especially, careful attention should be paid to the tension, not only as to uniformity, but also as to the total amount of tension placed on the wire. This should be just sufficient to keep the wire tight without stretching it enough to reduce its sectional area. The author has personally stretched No. 27 wire down to nearly No. 28 by simply adjusting the tension on the winding machine.

Square or rectangular windings require a very strong tension while being wound, otherwise the wire will not lie closely on the flat faces of the form or insulated core. With the round or elliptical types, the wire naturally tends to lie close on account of the curvature, as the wire tends to take the shortest path.

It is thus seen that the activity of the winding will depend largely upon the tension.

154. SQUEEZING

A method which is sometimes resorted to in practice, to increase the activity of the winding, is to squeeze the winding longitudinally, in order to crowd the turns as closely together as possible. It will be remembered that the vertical tension is greater than the lateral. While enameled and so-called bare-wire windings cannot be safely treated in this manner, silk and cotton covered wire windings may have their activity increased nearly 20 per cent in some cases, without injury to the insulation, or affecting the resistance of the winding, the degree of squeezing being dependent upon the ratio of insulation to copper.

In following this method it is customary to predetermine the amount of squeezing which may safely be applied, and then calculate the resistance or turns in the usual manner, assuming the increased length of the winding for L. After the winding is formed on the bobbin or on a tube of insulating material, it is squeezed to the length previously determined.

In one method a false core is placed end to end with the true core of the bobbin. After the winding has been squeezed, the false core will fall out. This is but one suggestion for the many ways in which this may be accomplished.

155. INSULATED WIRE WINDINGS WITH PAPER BETWEEN THE LAYERS

Electromagnets for use on lighting and power circuits require very carefully constructed windings. By placing thin sheets of paper between the layers of wire, the coils consist of smooth even layers well insulated from each other, and although the activity is slightly reduced, this type of winding has found great favor and is extensively used. A table, showing the values of N_a and ρ_v, for Enamelite Wires with one and two wraps of 1-mil paper between consecutive layers, will be found on page 313.

For this type of winding

$$d_1^2 = (d + i + s)(\mathrm{d} + i + P). \qquad (303)$$

The number of layers

$$n = \frac{T}{(d + i + P)}. \qquad (304)$$

and the turns per unit length

$$m = \frac{1}{(d + i + s)}. \qquad (305)$$

Wherein

$d =$ diameter of wire.

$i =$ increase due to insulation.

$s =$ lateral allowance between turns, edge to edge.

$P =$ vertical allowance, or thickness of paper.

In this case the paper should project for varying distances beyond the winding at each end of the coil, to prevent leakage or discharge from one layer to another, around the edge of the intervening paper.

156. DISK WINDING

While the disk winding is not extensively used on electromagnets, a description will not be out of place here.

One great disadvantage in the ordinary method of winding electromagnets is in the fact that the difference of potential between adjacent layers is liable to break down the insulation, as mentioned in Art. 129.

The disk winding is designed to overcome this difficulty, and is wound spirally, like the mainspring of a clock or watch. The conductor is usually in the form of a flat ribbon, and is insulated with silk, cotton, paper, or other insulating material. The difference of potential between successive layers is very slight since each layer consists of but one turn.

The disks are then placed side by side, connected together, electrically, and insulated from each other by mica disks. By this arrangement the total difference of potential is across the length of the winding instead of across the thickness of the winding, which is a great advantage. The space economy of this form of winding is also great, as there are no interstices between the wires outside of the insulation, as in the case of round-wire windings, and there is no pitch in the turns. In this latter connection the ideal condition is attained.

Sometimes round-wire windings are made in the form of disks, and connected, as just explained. This is necessary where fine wire is used, but in this case the turns incline toward the core slightly, as in the case of the regular round-wire winding.

157. CONTINUOUS RIBBON WINDING

In what may be called a modification of the disk winding, the ribbon is wound on edge around the insulated core, with suitable insulation placed between adjacent turns.

This winding consists of but one layer, and is adapted only for strong currents. It is very efficient with a high coefficient of ψ, and also possesses a high coefficient of heat conductivity, owing to the continuity of the copper from the inside to the outside of the coil.

158. MULTIPLE-WIRE WINDINGS

Windings for various electrical purposes often consist of several wires wound simultaneously, the wires form-

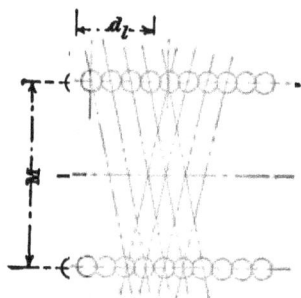

FIG. 201. — Four-wire Winding.

ing separate circuits, or with their terminals connected together to act as one conductor. When the wires are grouped as one circular strand, the winding is more effective than when the wires are wound on side by side in the form of a ribbon, owing to the greater pitch in the latter case. (See Fig. 201.)

When calculating windings of the class, it is important to use formula (220), p. 236, for the turns, and determine the resistance from the turns thus deduced. The turns and resistance may be calculated for each wire separately, or the total resistance may be determined from the total number of turns.

159. DIFFERENTIAL WINDING

This winding consists of two similarly insulated wires wound simultaneously side by side. Only one of the wires is used for exciting purposes, however, the other wire having its inside and outside ends connected together, thus forming a closed circuit.

It is extensively used where sparking, due to self-inductance, is detrimental to the contacts of the controlling device. When the circuit of the exciting winding is opened, the short-circuited winding absorbs the magnetic energy which would, otherwise, cause a momentary high voltage in the exciting circuit.

This winding is calculated by the methods described in Art. 158. Owing to the fact that only one half of the total winding space is available for the exciting coil, it is very inefficient.

160. ONE COIL WOUND DIRECTLY OVER THE OTHER

When both coils have the same size of insulated wire, simply calculate as one coil, making due allowance for the insulation between the two windings.

When the sizes of wire are different, the coils will, of course, have to be calculated separately, using the outer dimension of the first coil plus insulation for the inner dimension of the outer coil.

161. WINDING CONSISTING OF TWO SIZES OF COPPER WIRE IN SERIES

It was stated that the diameter of wire calculated for a given resistance and number of turns in a fixed winding space usually falls between two standard sizes of wire.

When a special diameter of wire is not obtainable or desirable, and the calculated diameter of wire is not sufficiently near a standard size to warrant the use of the latter, two wires may be employed, the average resistance and turns of which will be approximately as desired.

Since the number of turns will be inversely proportional to the difference between the required resistance and the resistance which would be obtained by using the adjacent sizes,

$$N_1 = \frac{N(\rho_{v\,max} - \rho_{vcal})}{\rho_{v\,max} - \rho_{v\,min}}, \qquad (306)$$

and

$$N_2 = \frac{N(\rho_{v\,cal} - \rho_{v\,min})}{\rho_{v\,max} - \rho_{v\,min}}, \qquad (307)$$

where N_1 = number of turns of smaller wire,

and N_2 = number of turns of larger wire.

Then, $$N = N_1 + N_2. \qquad (308)$$

The thickness or depth of the winding for each size of wire will be

$$T_1 = \frac{T(\rho_{v\,max} - \rho_{v\,cal})}{\rho_{v\,max} - \rho_{v\,min}}, \qquad (309)$$

$$T_2 = \frac{T(\rho_{vcal} - \rho_{v\,min})}{\rho_{v\,max} - \rho_{v\,min}}, \qquad (310)$$

where T_1 = thickness of smaller wire winding,

and T_2 = thickness of larger wire winding.

Then, $$T = T_1 + T_2. \qquad (311)$$

As the smaller winding will have the greater loss in watts, and therefore become the hotter of the two, it is customary to place it over the larger wire winding.

It is then only necessary to determine the value of the average perimeter p_a, for each winding, to calculate the resistance, the length of the winding being constant.

Sometimes it is required to obtain a high resistance in an electromagnetic winding which has so small a winding space that even No. 40 copper wire will not produce the required resistance. In such cases the smallest available copper wire should be used, the balance of the resistance being obtained with some resistance wire. The above method is applicable to this case.

162. RESISTANCE COILS

These are calculated after the same manner as electromagnetic windings consisting of copper wire, but the resistance will vary in direct proportion to the relative resistances of copper and resistance wire. Thus, if a winding is to contain 10,000 ohms of Climax wire, divide by 50 and calculate the same as though the coil was to consist of copper; that is, the same as if the resistance was to be $\dfrac{10,000}{50} = 200$ ohms, using the regular copper wire charts.

Resistance coils are usually wound *Non-inductively;* that is, two wires are connected together, the joint being thoroughly insulated, and fastened to one of the heads of the spool, and then the two wires are wound on together in parallel. Both wires, therefore, have the same number of turns and the current flows toward the inner connection in an opposite direction to which it flows out, thus neutralizing any magnetic tendency, and eliminating the inductive effects.

163. MULTIPLE-COIL WINDINGS

Among the several methods of winding electromagnets so that the sparking due to self-induction shall be minimized, the best two are what are known as the *Differential* winding and the *Multiple-coil* winding. This does not refer to external methods of compensation.

The differential winding is less expensive than the multiple-coil winding, for a given voltage, as much coarser wire and less turns are used in the former than in the latter.

The extreme commercial condition for the multiple-coil winding is illustrated in Fig. 202, where the respective ends of each layer are all connected together. Since in this type of winding the full voltage is across each of the multiple wires or separate coils, this arrangement is not practicable for comparatively high voltages on coils of moderate size, as the wire in each layer would have to resist the full-line voltage without overheating. This applies most fully to the first or inner coil, which, having the least resistance, must pass the most current.

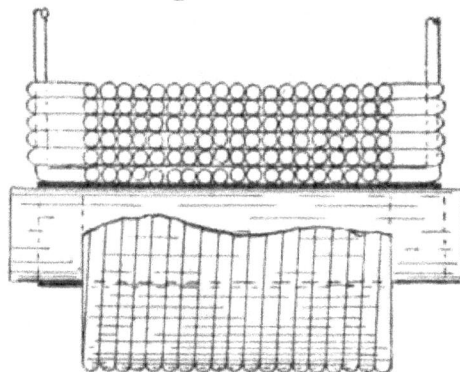

FIG. 202. — Winding with Layers connected in Multiple.

The principle of the multiple-coil winding is that the inner turns (or separate coils) have less resistance but a greater coefficient of self-induction than the outer turns, owing to the different diameters of the inner and outer turns, and hence the time-constants of the separate windings are different.

For electromagnets three or four inches in length by two or three inches in diameter, to operate on 110-volt, direct-current circuit, the author has had excellent results from six separate windings per spool, wound over each other, and connected in multiple, as in Fig.

203.   While Fig. 203 shows the terminals at opposite ends of the bobbin, by making the layers even in num-

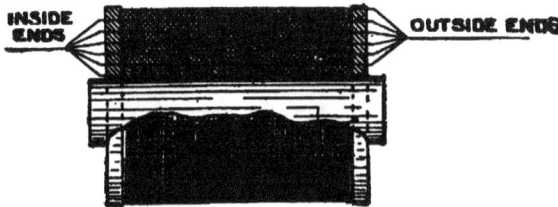

FIG. 203. — Practical Multiple-coil Winding.

ber, the terminals may be brought out at the same end of the winding, as in Fig. 204.   This arrangement produces the same general effect as six layers of coarser wire, with a corresponding decrease in voltage, and, therefore, is just as efficient, with the exception that there will be less copper in the fine-wire winding, owing to the greater ratio of insulating material on the fine wire.

In order to ascertain the safe current carrying capacity of the winding, a plain or regular winding may be assumed.   As an example consider the bobbin in Fig. 205. The winding is to be connected directly across 110 volts direct current, and will be in use at such intervals that 13.75

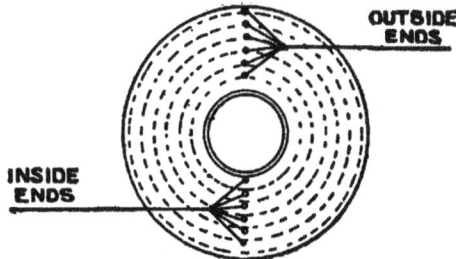

FIG. 204. — Method of bringing out Terminals.

watts will be permissible for the winding.   The resistance of the winding will then be

$$R = \frac{E^2}{P} \ (312), \ = \frac{12,100}{13.75} = 880 \text{ ohms.}$$

Referring to Fig. 206, M is the mean or average diameter of the entire winding space, and M_a, M_b,

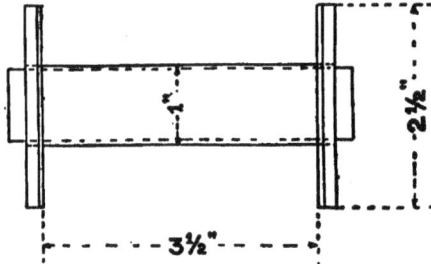

FIG. 205. — Bobbin.

etc., are the average diameters of the coils constituting the multiple-coil winding. Therefore, M also represents the average of the mean diameters of the coils a, b, c, etc.

If all the coils were of the same diameter, and hence of the same resistance, the joint resistance would be proportional to $\dfrac{M}{n}$, where n is the number of coils, 6 in this case.

Although the mean diameters M_a, M_b, etc., are variable, they vary in a direct ratio to one another, and therefore by comparing $\dfrac{1}{n}$ of their "average mean diameter" with their "joint mean diameter," we may obtain a basis from which to compute the proper wire to use which shall give the desired resistance when the coils are connected in multiple. $\dfrac{1}{n}$ of the average mean diameter $= \dfrac{M}{n} = \dfrac{1.75}{6} = .292 = M_a$ (see Fig. 206). The joint mean diameter will be

$$M_j = \cfrac{1}{\dfrac{1}{M_a} + \dfrac{1}{M_b} + \dfrac{1}{M_c} + \dfrac{1}{M_d} + \dfrac{1}{M_e} + \dfrac{1}{M_f}} \tag{313}$$

$$= \frac{1}{.889 + .727 + .615 + .533 + .471 + .421} = \frac{1}{3.656} = .274.$$

Therefore, the ratio will be $.274 : .292 = .94 : 1$, which means that the joint resistance will be $.94$ times

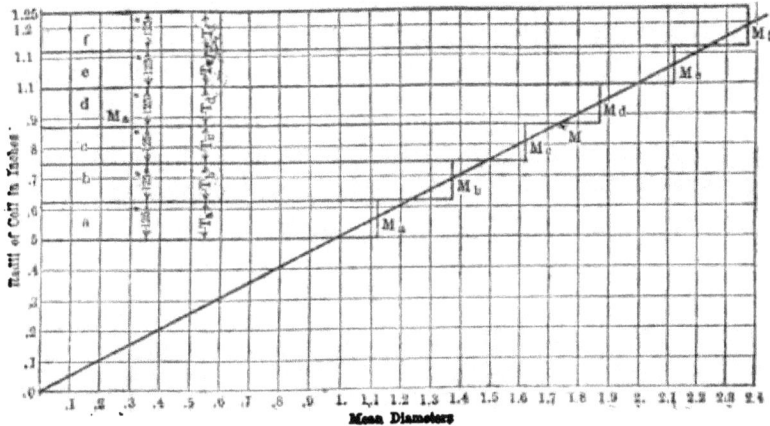

FIG. 206. — Mean Diameters of Multiple-coil Windings.

$\frac{1}{n}$ of the average resistance of the coils, and since the latter is

$$R_A = \frac{R_a + R_b + R_c + R_d + R_e + R_f}{n^2}, \qquad (314)$$

the joint resistance

$$R_j = \frac{M_j (R_a + R_b + R_c + R_d + R_e + R_f)}{M_a \, n^2} \qquad (315)$$

$$= \frac{.94 \, (R_a + R_b + R_c + R_d + R_e + R_f)}{n^2},$$

and since $R_j = 880$ and $n = 6$,

$$(R_a + R_b + R_c + R_d + R_e + R_f) = \frac{880 \times 36}{.94} = 33,700.$$

Therefore, the ohms per cubic inch $= \dfrac{33,700}{14.43} = 2335$,

which corresponds very nearly to No. 39 B. & S. wire with 2-mil insulation.

If paper is inserted between the coils, and an insulating varnish used, a No. 39 B. & S. wire with 1.5-mil silk insulation would meet this requirement. Therefore, assuming the ohms per cubic inch to be 2335, and calculating the volumes of the coils separately, which are found to be 1.55, 1.89, 2.23, 2.58, 2.92, and 3.26 cubic inches, respectively, the resistances will be 3620, 4410, 5210, 6020, 6820, and 7610 ohms, respectively. Their joint resistance will then be:

$$\frac{1}{\frac{1}{3620} + \frac{1}{4410} + \frac{1}{5210} + \frac{1}{6020} + \frac{1}{6820} + \frac{1}{7610}}$$

$$= \frac{1}{.000276 + .000226 + .000192 + .000166 + .000147 + .000131}$$

$$= \frac{1}{.001138} = 880.$$

Now also the ratio $\dfrac{M_j}{M_a} = \dfrac{M_j}{\left(\dfrac{M}{n}\right)} = \dfrac{M_j n}{M}$

$$= \frac{n}{M\left(\dfrac{1}{M_a} + \dfrac{1}{M_b} + \dfrac{1}{M_c} + \dfrac{1}{M_d} + \dfrac{1}{M_e} + \dfrac{1}{M_f}\right)}, \qquad (316)$$

and since $(R_a + R_b + R_c + R_d + R_e + R_f) = R_s,$ (317)

$$R_j = \frac{n}{M\left(\dfrac{1}{M_a} + \dfrac{1}{M_b} + \dfrac{1}{M_c} + \dfrac{1}{M_d} + \dfrac{1}{M_e} + \dfrac{1}{M_f}\right)} \times \frac{R_s}{n^2}$$

$$= \frac{R_s}{M_n\left(\dfrac{1}{M_a} + \dfrac{1}{M_b} + \dfrac{1}{M_c} + \dfrac{1}{M_d} + \dfrac{1}{M_e} + \dfrac{1}{M_f}\right)}; \qquad (318)$$

and hence

$$R_s = R_j M_n\left(\dfrac{1}{M_a} + \dfrac{1}{M_b} + \dfrac{1}{M_c} + \dfrac{1}{M_d} + \dfrac{1}{M_e} + \dfrac{1}{M_f}\right). \qquad (319)$$

Therefore, to calculate the ohms per cubic inch, ρ_v, direct, use the formula

$$\rho_v = \frac{R_j M_n \left(\dfrac{1}{M_a} + \dfrac{1}{M_b} + \dfrac{1}{M_c} + \dfrac{1}{M_d} + \dfrac{1}{M_e} + \dfrac{1}{M_f} \right)}{V}, \quad (320)$$

where V is the total volume of the winding space in cubic inches.

In practice, paper or other insulating material is placed between the separate coils, since the total voltage is between adjacent coils, and hence the space occupied by the paper or other insulating material must be deducted from the total winding volume. It will be observed that the resistance of the inner coil a is only 3620 ohms, while that of the outer coil f is 7610 ohms, which is more than twice the resistance of coil a. Therefore, there will be generated in the inner coil a twice as much heat as generated in the 'outer coil, and hence if the proper resistance is not provided, coil a will get very hot unless there is a sufficient mass of core material to conduct away the heat to be radiated from the frame.

Instead of calculating the " ohms per cubic inch " for the wire used in the multiple-coil winding, a regular winding of one coil may be assumed, and the diameter of the wire, found by comparing the ohms per cubic inch with the diameter of the wire in Figs. 216 to 221.

The diameter of the wire for the multiple-coil winding will then be

$$d_s = \frac{d}{\sqrt{n}}, \quad (321)$$

wherein n is the number of coils.

As this is only an approximate method, it is best to assume 4-mil insulation for the regular winding, and 2-mil or 1.5-mil insulation for the multiple-coil winding. To determine the proper size of wire (4-mil increase insulation in this case) the ohms per cubic inch must now be found, which of course are equal to the resistance divided by the volume of the winding in cubic inches.

The volume of the winding will be πMLT. In the case considered

$$V = 3.1416 \times \left(\frac{2.5 + 1}{2}\right) \times 3.5 \times \left(\frac{2.5 - 1}{2}\right) =$$

$3.1416 \times 1.75 \times 3.5 \times .75 = 14.43$ cubic inches, and the ohms per cubic inch will be $\frac{880}{14.43} = 61$. A glance at Fig. 219 shows that the nearest B. & S. copper wire with 4-mil insulation is No. 31.

Formula (321) is derived from the fact that in any case the resistance per unit volume for any wire is inversely proportional to the fourth power of the diameter of the wire. (The presence of the insulation varies this somewhat, so formula (321) is only approximate.) And since $R_s = n^2 R$ (approximately), $d_s^4 = \frac{d^4}{n^2}$, where $d_s = \frac{d}{\sqrt{n}}$.

164. RELATION BETWEEN ONE COIL OF LARGE DIAMETER AND TWO COILS OF SMALLER DIAMETER, SAME AMOUNT OF INSULATED WIRE WITH SAME DIAMETER AND LENGTH OF CORE IN EACH CASE

In order to make the relation clear assume an actual example: In both cases assume $\frac{1}{2}$ inch diameter cores of the same length. Assume the diameter of the insulating sleeve D_1 to be .55 inch, and the length of the winding L to be 2 inches in each case. The total resistance in each case will be 100 ohms, the wire No. 28 S. S. C. ($\rho_v = 19.5$), and the e. m. f., say 50 volts.

Case 1. One coil only, of 100 ohms.

By formula (276),

$$D = \sqrt{\frac{1.273\,R}{\rho_v\,L} + D_1{}^2} = \sqrt{3.56} = 1.89.$$

From (267), $M = \dfrac{D + D_1}{2} = 1.22.$

Therefore, $p_a = \pi M = 3.83$, and from (226)

$$IN = \frac{E}{\rho_l p_a} = 2420.$$

Case 2. Two coils of 50 ohms each.

From (276) $D = \sqrt{\dfrac{1.273\,R}{\rho_v\,L} + D_1{}^2} = \sqrt{1.93} = 1.39.$

From (267) $M = \dfrac{D + D_1}{2} = .97.$ Therefore, $p_a = 3.04.$

In this case $E = 25$ for each coil. Therefore, $IN = \dfrac{E}{\rho_l p_a} = 1525$ for each coil, or 3050 ampere-turns for two coils, making an increase of 26 per cent for case 2 over case 1, with the same kind and amount of insulated wire in each case.

165. Different Sizes of Windings connected in Series

When two windings of different volumes are to be connected in series, each being wound with the same size of wire, and with a fixed total resistance, it is customary to proceed as follows:

EXAMPLE: Two windings, when connected in series, are to have a total resistance of 50 ohms. Their relative volumes are as $1:6$. What should be the resistance of each?

SOLUTION: Let $R_1 =$ resistance of first winding.

$R_2 =$ resistance of second winding.

$R_1 + R_2 = 50$, and since $6 R_1 = R_2$, $6 R_1 + R_1 = 50$, whence, $7 R_1 = 50$, or $R_1 = 7.143$. Since $R_2 = 6 R_1$, $R_2 = 42.858$.

166. SERIES AND PARALLEL CONNECTIONS

When several electromagnets are to be operated simultaneously, they may be connected in two different ways; that is, in series or in parallel. The former method is the cheaper, as coarser wires may be used in the winding. The total current consumption, however, will be about the same in both cases. The multiple arrangement is the safer, however, as any of the connections to the electromagnet may be broken without affecting the rest of the electromagnets in the line, while if any of the connections should become broken in the series arrangement, the entire circuit would become inoperative. On the other hand, there is more danger from short-circuits in the multiple than in the series arrangement.

Where the electromagnets are connected in series, the total line current passes through all of the windings, while each winding consumes but a portion of the total voltage, whereas, in the multiple arrangement, the total line voltage is across the terminals of each winding, while each winding consumes but a portion of the total current; therefore, the multiple arrangement requires much finer wire in the windings of the electromagnets. The cost of operating, however, will be about the same in both cases, as the only variation will be in the relation of insulation to copper in the finer or coarser insulated wires. The above holds true where the line has a negligible resistance.

167. WINDING IN SERIES WITH RESISTANCE

There are many cases in practice where an electromagnet is operated in a circuit containing a resistance in series with, and external to, the resistance of the winding itself. Theoretically there will always be some external resistance, as, for instance, the resistance of the leads to the electromagnet and other wiring; but since the resistance of the windings for local use, and particularly if designed to remain in circuit indefinitely without overheating, is great as compared with the resistance of the wiring and source of energy, this external resistance is not usually considered.

However, in this article the resistance in the circuit external to the resistance of the winding will be taken into consideration. The old rule "Make the internal and external resistances equal" holds for the maximum electrical power in watts which may be obtained in a winding, and also for the maximum magnetizing force for the winding under certain conditions; but this rule is not strictly correct as applied to the conditions in actual practice. Under these conditions the winding of the magnet should have slightly less resistance than the line, in order to do the most work, providing, of course, that the winding volume is great enough to prevent the winding from becoming overheated.

With fixed winding volume, the activity will vary with the size of the insulated wire. Hence, if the resistance of the winding be increased, the activity will be decreased.

If we let E = voltage of source of energy,

E_1 = voltage across winding,

R = resistance of winding,

and R_1 = all other resistance in the circuit,

then

$$E_1 = \frac{RE}{R + R_1}. \tag{322}$$

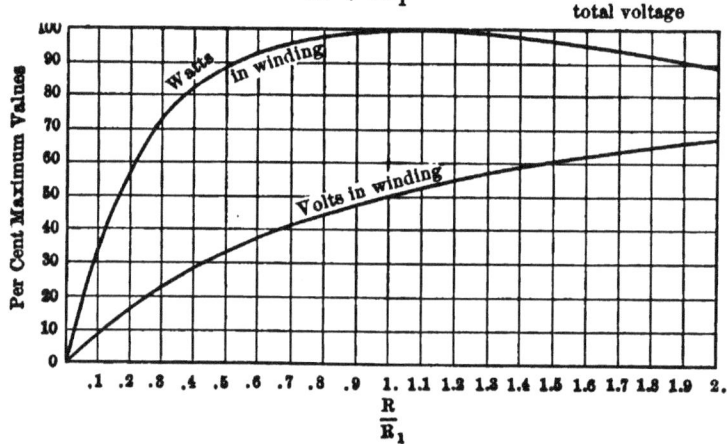

FIG. 207. — Characteristics of Two Resistances in Series.

Figure 207 shows the percentage of maximum values for volts and watts for the winding according to this rule, the watts being maximum when

$$R = R_1, \text{ and } E_1 = \frac{E}{2}.$$

Now consider a magnet winding of fixed dimensions in series with an external resistance, equal to that of the magnet winding, and a source of energy of constant voltage. When $\frac{d^2}{d_1^2} = 1$, which would be the ideal condition for round wire, the ampere-turns will attain their maximum value for that size of wire, but if, as in the case of a No. 30 wire insulated with 4-mil increase double silk $\left(\frac{d^2}{d_1{}^2} = 0.51\right)$, it would be necessary to use a wire of approximately twice the resistance per unit length in order to keep the resistance of the magnet winding the same in the latter case as in the former, and hence when the watts in the magnet winding were maximum, the ampere-turns would be approximately 50 per cent of their maximum value for ampere-turns in the ideal case.

However, there are two distinct effects to be considered: (a) in which case the space occupied by the insulation on the wire bears a constant relation to the space occupied by the wire throughout the various sizes of wire, and (b) in which the thickness of insulating material on the wire is constant for all the various sizes of wire. In the former case (a) it would be necessary to use a different thickness of insulating material for each size of wire, in order to keep $\frac{d^2}{d_1{}^2}$ constant, and hence, case (b) is the method adopted in practice. In this latter case the value of $\frac{d^2}{d_1{}^2}$ changes with every change in the size of the wire.

By assuming $E = 100$ and $R_1 = 100$, the relations in Figs. 208 to 210 have been calculated. Under these conditions the maximum value for the watts in the winding will be 25, when $E_1 = 50$ and $R = 100$. Figure 208 shows the relations for several values of $\frac{d^2}{d_1{}^2}$ under case (a) in which these values are constant. It will be noted that the ampere-turns and watts are maximum simultaneously for each respective value of $\frac{d^2}{d_1^2}$, and that this point is determined by the intersection of the volts curve with the ordinate 0.5. While the maximum value for watts is 25 for any size of wire, under these conditions, the ampere-turns vary in the ratio shown in Fig. 209.

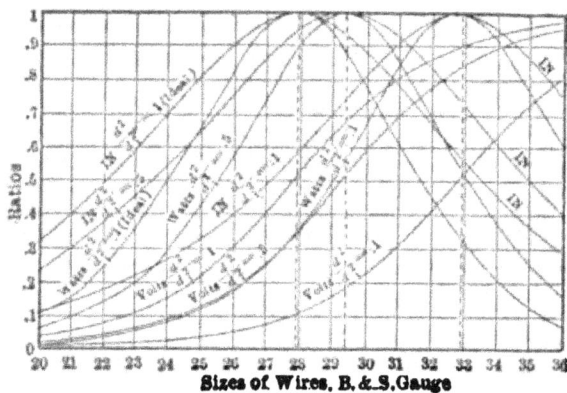

FIG. 208. — Effect with Variable Thickness of Insulation, $\frac{d^2}{d_1^2}$ Constant.

In Figs. 208 and 209 the relations expressed by the rule first referred to hold true, but in Fig. 210, where the values of $\frac{d^2}{d_1^2}$ are variable (case b), the watts and ampere-turns are not maximum simultaneously, although the point of maximum watts is determined by the intersection of the volts curve with the ordinate 0.5, as in case (a), Fig. 208.

The point for maximum ampere-turns with $\frac{d^2}{d_1^2}$ variable (Fig. 210) is determined by the intersection of the insulation curve with the ampere-turns curve for that

insulation. This gives the value of $\dfrac{d^2}{d_1{}^2}$ which will
show the percentage of ideal ampere-turns, as in
Fig. 209. This curve is reproduced in Fig. 210
(marked e), and its origin is at the intersection of the

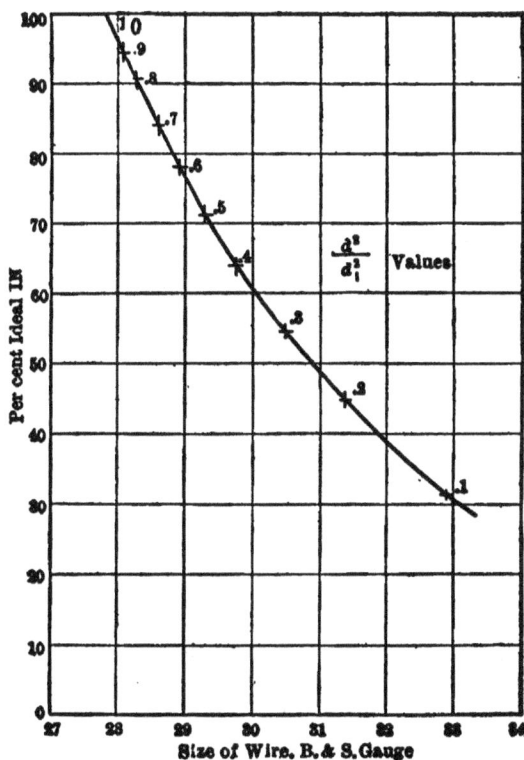

FIG. 209.— Effect of Insulation.

ideal volts curve and the ordinate 0.5, which ordinate
represents the maximum value for the percentage of
ideal IN (curve e).

The size of wire which will give the maximum
ampere-turns under these conditions, and with a given
thickness of insulation, is determined by the intersec-

tion of the volts curve for the given insulation with curve e. Having once plotted curve e, it is an easy matter to find the size of wire to give the maximum ampere-turns for any case, by first calculating the size of wire (assuming the insulation to be nil) which will produce the maximum ampere-turns for the ideal case $\left(\dfrac{d^2}{d_1{}^2} = 1\right)$. Two points for volts may then be cal-culated, taking into consideration the insulation on the

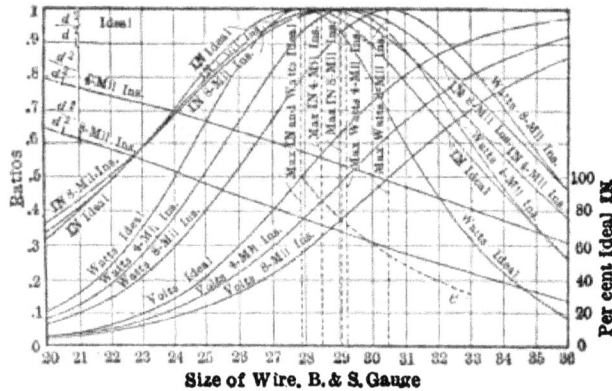

Fig. 210. — Effect with Constant Thickness of Insulation, $\dfrac{d^2}{d_1{}^2}$ Variable.

wire. One of these points should be taken on the ordinate 0.5, $i.e.$ when $R = R_1$ $\left(\text{and hence, } E_1 = \dfrac{E}{2}\right)$, calculating the size of the wire. The other point should be taken for voltage on the abscissa represent-ing the size of wire to produce ideal ampere-turns, re-ferred to above. Connecting these points will locate the size of wire on curve e, which will produce the maximum ampere-turns with the given insulation as previously explained.

Figure 211 shows this principle, in which curve e is

plotted to the same scale as in Fig. 210. In this case, however, the curve is a straight line at an angle of 50° with the ordinate 0.5.

As an example, assume a solenoid with an available winding volume of 20 cubic inches, and with an average diameter of 2 inches to be operated on a 220-volt circuit, in series with a resistance of 200 ohms. Assuming $\frac{d^2}{d_1^2} = 1$, and by rearranging (263), the diameter of the wire is found to be 0.017 inch, or between Nos. 25 and 26 (approximately No. 25.4) B. & S. By (226), or referring to Fig. 188, p. 247, the ampere-turns are found to be approximately 5850. Now when $E = E_1$, $R = R_1$, and hence the resistance of the solenoid, when watts are maximum, will be 200 ohms. Assuming 8-mil insulation, and calculating d from (263), the diameter of the wire is found to be 0.0136 inch, or between Nos. 27 and 28, or by formula (190), the fractional size is found to be very near No. 27.5.

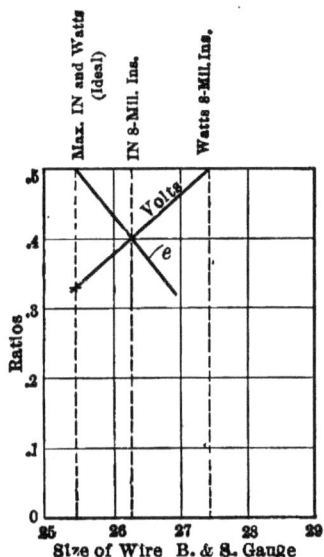

FIG. 211.—Curve "e" as a Straight Line.

The point where the voltage curve intersects the ideal abscissa is calculated from the same size of wire as in the ideal case, i.e. No. 25.4 or 0.017 inch diameter, but with 8-mil insulation. By (273) the resistance is found to be 95.5 ohms, and by (322) the voltage E_1 is 71, or 32.2 per cent of 220 volts. Therefore the size of wire with 8-mil insulation, which will produce the maximum ampere-turns under these conditions, is approximately No. 26¼.

As a practical proposition, the size of wire may be calculated for a 50 per cent drop in volts across the winding, using the next larger size of wire, unless the value of $\dfrac{d^2}{d_1^2}$ is near unity. Therefore, calculate the size of wire to use, assuming the resistance of the coil to be equal to the total external resistance, and then try the next larger size of wire, selecting that which gives the greatest number of ampere-turns.

Now
$$E_1 = \frac{RE}{R+R_1}, \tag{322}$$

and
$$R = \frac{cp_a LT}{d^2(d+i)^2}. \tag{323}$$

Substituting the value of R from (322) in (323),

$$E_1 = \frac{E}{\left(\dfrac{R_1 d^2(d+i)^2}{cp_a LT}\right) + 1}. \tag{324}$$

The ampere-turns are maximum when $\dfrac{E_1 d^2}{cp_a}$ is maximum.

Therefore,

$$IN = \frac{E}{\left(\dfrac{R_1(d+i)^2}{TL}\right) + \dfrac{cp_a}{d^2}}. \tag{325}$$

168. EFFECT OF POLARIZING BATTERY

When a battery is to be used for continuously and interruptedly operating an electromagnet of low electrical resistance, a non-polarizing type of battery, preferably a storage battery, should be used.

If, however, the electromagnet is of the horseshoe or plunger type, where very little current is required to maintain the required pull near the cores, and the operating current is to be left on for a considerable period of time, it is sometimes desirable to use a polarizing primary battery; as the current will fall off rapidly after the electromagnet has performed its duty, and, therefore, the winding will not become so heated as it would if the full strength of the battery current should pass through the winding. There will also be a saving in energy, thus prolonging the life of the battery.

This arrangement also permits of a smaller electromagnet being used than if the operating current were to be left on the winding continuously, thus saving in first cost also.

169. General Precautions

The success of accurately calculating electromagnetic windings depends upon close attention to details. The wire should always be carefully gauged in several places with a ratchet-stop micrometer, allowances being made for very small variations in the diameter. The diameter over the insulation should also be carefully observed.

The winding volume should be accurately determined, and the insertion of paper into the winding avoided as much as possible. The tension should be constant and not great enough to stretch the wire. The turns and resistance should be carefully compared, as this will aid in detecting any irregularities in the winding.

CHAPTER XX

HEATING OF ELECTROMAGNETIC WINDINGS

170. Heat Units

The C. G. S. unit of heat is the *Calorie*, and is the quantity of heat required to raise the temperature of one gram of water one degree C. at or near its temperature of maximum density, 4° C.

The *Mechanical Equivalent* is 4.16×10^7 ergs.

The unit of heat, in English measure, is the *British Thermal Unit*, abbreviated B. T. U., and is the quantity of heat which will raise the temperature of one pound of water one degree F. at or near its temperature of maximum density, 39.1° F.

The mechanical equivalent was found by Joule to be 772 foot-pounds. Thus 772 foot-pounds is called *Joule's Equivalent*. Professor Rowland, however, found the equivalent to be 778 foot-pounds. Hence, 1 B. T. U. = 778 foot-pounds.

$$1 \text{ foot-pound} = \frac{1}{778} = 0.001285 \text{ B. T. U.}$$

One calorie = 0.00396 B. T. U.; 1 B. T. U. = 251.9 calories.

The electrical unit of heat is the *Joule* or *Watt-second*, and is the quantity of heat generated in one second by one watt of energy. One joule = 10^7 ergs.

171. Specific Heat

The *Specific Heat* of a body at any temperature is the ratio of the quantity of heat required to raise the temperature of the body one degree, to the quantity of heat required to raise an equal mass of water at or near its temperature of maximum density, through one degree.

The specific heat of copper at 50° C. or 122° F. is 0.0923, and for German silver, at the same temperature, 0.0947.

172. THERMOMETER SCALES

The standard thermometer scales in common use are the *Fahrenheit* and *Centigrade*.   In the former, the tem-

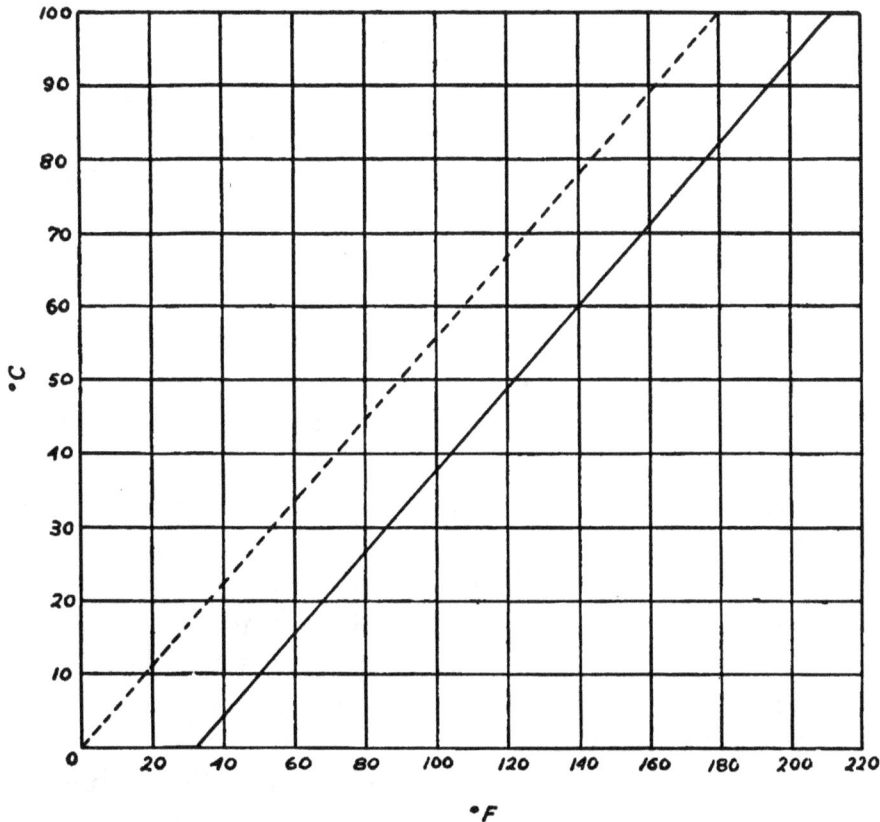

FIG. 212. — Comparison of Thermometer Scales.

perature of melting ice is marked 32° and the temperature of boiling water 212°. The centigrade scale,
invented by Calcius, is divided into 100 equal parts
between 0° for the freezing point, and 100° for the
boiling point ; hence its name. In both, the scales are
projected as far above the boiling point or below the
freezing point, as may be desired. The centigrade
scale is preferable.

Conversion from one scale to the other may be
accomplished by means of the following formulæ:

$$F° = \tfrac{9}{5}C° + 32. \qquad (326)$$

$$C° = \tfrac{5}{9}(F° - 32). \qquad (327)$$

Figure 212 also shows the relations.

The full line shows the scale relations, while the
dotted line shows the ratio between degrees, which is
as 5:9. This dotted line is to be used in converting
rise in temperature from one scale into the other.

173. HEATING EFFECT

An electric current flowing through a winding generates heat therein proportional to the watts lost in the
winding. If the winding consists of good heat-conducting material, and ample surface is provided for
the radiation of the heat, much more energy may be
applied than if the winding be poorly designed.

Much regarding the heat-resisting qualities and heat-
conducting properties of insulating materials was mentioned in Chapter XVII. It is obvious that heat may
be conducted through a thin winding much faster than
through a thick one.

Experience has shown that a coil of ordinary dimensions may remain in circuit continuously when the ap-

plied electrical power does not exceed 0.50 watt per square inch of superficial radiating surface.

Coils mounted on large iron cores which in turn are attached to the frames of machines have an advantage in the fact that the core conducts the heat away where it can be radiated rapidly.

174. Temperature Coefficient

Most of this article, as well as the data from which Fig. 213 was made, is taken from the Standardization Rules of the American Institute of Electrical Engineers.

The fundamental relation between the increase of resistance in copper and the rise of temperature may be taken as

$$R_t = R_o(1 + 0.0042\ t), \tag{328}$$

where R_o is the resistance at $t°$ C. of the copper conductor at $0°$ C., and R_t is the corresponding resistance.

This is equivalent to taking a temperature coefficient of 0.42 per cent per degree C. temperature rise above $0°$ C. For initial temperatures other than $0°$ C., a similar formula may be used, substituting the coefficients in Fig. 213 corresponding to the actual temperature. The formula thus becomes at $25°$ C.,

$$R_{i+r} = R_i\left(1 + \frac{0.3801\ r}{100}\right), \tag{329}$$

where R_i is the initial resistance at $25°$ C., R_{i+r} the final resistance, and r the temperature rise above $25°$ C.

In order to find the temperature rise in degrees C. from the initial resistance R_i at the initial temperature $i°$ C., and the final resistance R_{i+r}, use the formula

$$r = (238.1 + i)\left(\frac{R_{i+r}}{R_i} - 1\right). \tag{330}$$

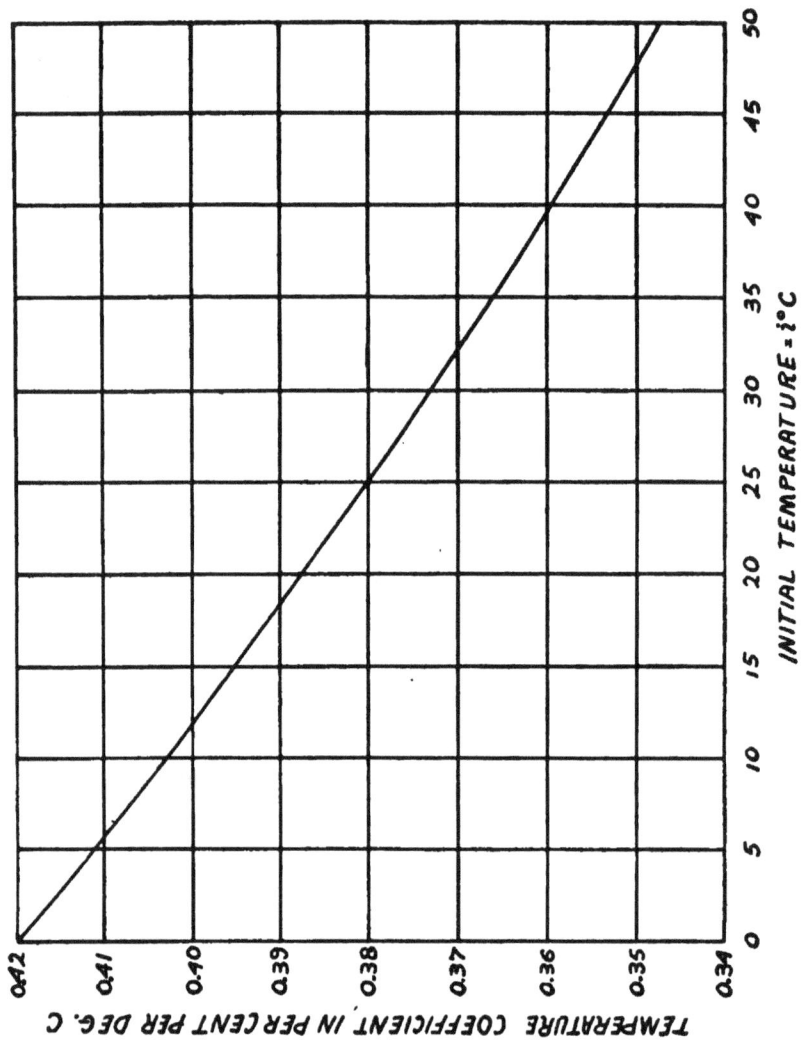

Fig. 213. — Temperature Coefficients.

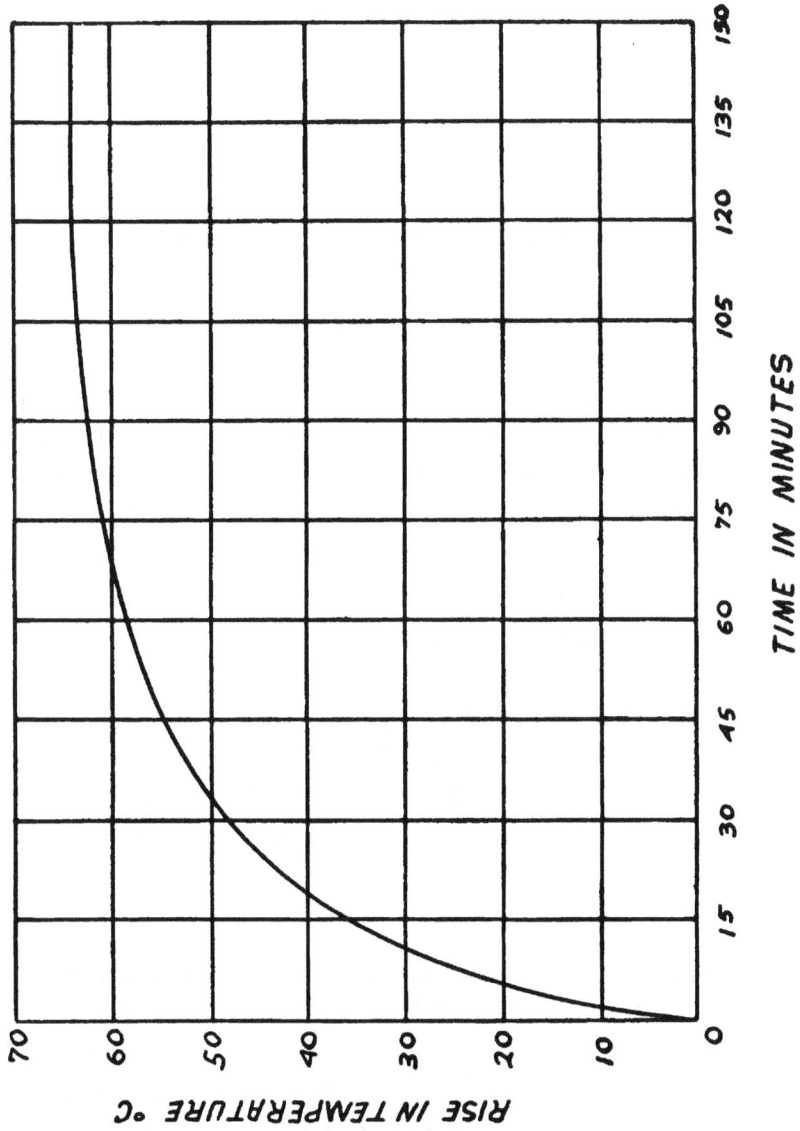

TIME IN MINUTES

Fig. 214. — Heat Test.

RISE IN TEMPERATURE °C

The amount of applied energy depends upon the place where the coil is to be used.

The resistance of a winding at the limiting temperature will, therefore, be

$$R_{i+r} = \frac{W}{S_r}, \qquad (331)$$

wherein W = total energy in coil, and S_r is the radiating surface.

175. Heat Tests

Experimental data for any particular coil may be obtained by placing *thermal coils*, which consist of a few turns of small wire, at different points in the winding. The rise in resistance of these coils will determine the rise in temperature of the winding, observations being taken from time to time, from which data a curve may be plotted. Such a curve is shown in Fig. 214, which is the result of a test of the small iron-clad solenoid of dimensions $L = 4.6$ cm., $r_a = 1.3$ cm. (See p. 108.) The winding should be designed for a rise in temperature considerably lower than that shown in the illustration, when silk or cotton insulation is used.

176. Activity and Heating

What was said in Art. 132 applies particularly to coils which are to be left in circuit indefinitely.

The greater the activity of the winding, the less will be the energy required, and, consequently, the less will be the heating, for an electromagnet of given dimensions.

Electromagnets for operating trolley signals sometimes have a resistance of several thousand ohms, depending upon the time they are to remain in circuit.

As a matter of fact, it is best to make magnets of this character "Fool-proof," by so designing the winding that the current may be left on continuously without overheating the coil.

The resistance of a winding which is to be left in circuit continuously will vary with the radiating surface.

For a very large magnet the resistance may be only a few hundred ohms, while for a small one it may be several thousand ohms. The figures given apply, of course, to high-voltage apparatus.

A magnet which is to be left in circuit indefinitely must, necessarily, be larger than one which is to exert the same pull through the same distance, but to only remain in circuit for a short time and then remain idle.

A large electromagnet has a greater winding volume than a small one; hence, its resistance may be much greater without greatly increasing the average perimeter and, therefore, without greatly reducing the ampere-turns.

By slightly increasing the thickness of the winding the resistance will be correspondingly increased, and the heating reduced without greatly reducing the ampere-turns for the same size of wire.

A winding completely enclosed in an iron shell may, if the space between the winding and shell be properly filled with a good heat-conducting insulating compound, have approximately 50 per cent more energy applied to it per unit surface area than if the coil were exposed to the open air. This will, of course, depend upon the radiating surface of the shell through which the heat is conducted from the winding.

CHAPTER XXI

TABLES AND CHARTS

The following tables and charts have been placed in a separate chapter in order to make them easily accessible for reference. The factors given for insulated wires are those which have been found to give the best results in practice. The weights of insulated wire are, however, for perfectly dry insulation, and in practice they may appear too low, owing to the hygroscopic properties of the insulating material.

The factors are expressed in English units. Thus,

$d =$ diameter of bare wire in inch.

$d_1 =$ diameter of insulated wire in inch.

$\rho_l =$ ohms per inch.

$\rho_w =$ ohms per pound for bare wires.

$\rho_v =$ ohms per cubic inch for insulated wires.

$W_v =$ pounds per cubic inch for insulated wires.

$N_a =$ turns per square inch.

The temperature for which these tables and charts have been calculated is 20° C. or 68° F.

STANDARD COPPER WIRE TABLE *

Giving weights, lengths, and resistances of wires at 20° C. or 68° F., of Matthiessen's Standard Conductivity, for A. W. G. (Brown & Sharpe).

A.W.G.	DIAMETER	AREA	WEIGHT		LENGTH		RESISTANCE	
B. & S.	Inches	Circular Mils	Lbs. per Ft.	Lbs. per Ohm	Feet per Lb.	Feet per Ohm	Ohms per Lb.	Ohms per Ft.
0000	.460	211,600	.6405	13,090	1.561	20,440	.00007639	.00004893
000	.4096	167,800	.5080	8,232	1.969	16,210	.0001215	.00006170
00	.3648	133,100	.4028	5,177	2.482	12,850	.0001931	.00007780
0	.3249	105,500	.3195	3,256	3.130	10,190	.0003071	.00009811
1	.2893	83,690	.2533	2,048	3.947	8,083	.0004883	.0001237
2	.2576	66,370	.2009	1,288	4.977	6,410	.0007765	.0001560
3	.2294	52,630	.1593	810.0	6.276	5,084	.001235	.0001967
4	.2043	41,740	.1264	509.4	7.914	4,031	.001963	.0002480
5	.1819	33,100	.1002	320.4	9.980	3,197	.003122	.0003128
6	.1620	26,250	.07946	201.5	12.58	2,535	.004963	.0003944
7	.1443	20,820	.06302	126.7	15.87	2,011	.007892	.0004973
8	.1285	16,510	.04998	79.69	20.01	1,595	.01255	.0006271
9	.1144	13,090	.03963	50.12	25.23	1,265	.01995	.0007908
10	.1019	10,380	.03143	31.52	31.82	1,003	.03173	.0009972
11	.09074	8,234	.02493	19.82	40.12	795.3	.05045	.001257
12	.08081	6,530	.01977	12.47	50.59	630.7	.08022	.001586
13	.07196	5,178	.01568	7.840	63.79	500.1	.1276	.001999
14	.06408	4,107	.01243	4.931	80.44	396.6	.2028	.002521
15	.05707	3,257	.009858	3.101	101.4	314.5	.3225	.003179
16	.05082	2,583	.007818	1.950	127.9	249.4	.5128	.004009
17	.04526	2,048	.006200	1.226	161.3	197.8	.8153	.005055
18	.04030	1,624	.004917	.7713	203.4	156.9	1.296	.006374
19	.03589	1,288	.003899	.4851	256.5	124.4	2.061	.008038
20	.03196	1,022	.003092	.3051	323.4	98.66	3.278	.01014
21	.02846	810.1	.002452	.1919	407.8	78.24	5.212	.01278
22	.02535	642.4	.001945	.1207	514.2	62.05	8.287	.01612
23	.02257	509.5	.001542	.07589	648.4	49.21	13.18	.02032
24	.02010	404.0	.001223	.04773	817.6	39.02	20.95	.02563
25	.01790	320.4	.0009699	.03002	1,031	30.95	33.32	.03231
26	.01594	254.1	.0007692	.01888	1,300	24.54	52.97	.04075
27	.0142	201.5	.0006100	.01187	1,639	19.46	84.23	.05138
28	.01264	159.8	.0004837	.007466	2,067	15.43	133.9	.06479
29	.01126	126.7	.0003836	.004696	2,607	12.24	213.0	.08170
30	.01003	100.5	.0003042	.002953	3,287	9.707	338.6	.1030
31	.008928	79.70	.0002413	.001857	4,145	7.698	538.4	.1299
32	.007950	63.21	.0001913	.001168	5,227	6.105	856.2	.1638
33	.007080	50.13	.0001517	.0007346	6,591	4.841	1,361	.2066
34	.006305	39.75	.0001203	.0004620	8,311	3.839	2,165	.2605
35	.005615	31.52	.00009543	.0002905	10,480	3.045	3,441	.3284
36	.0050	25.0	.00007568	.0001827	13,210	2.414	5,473	.4142
37	.004453	19.83	.00006001	.0001149	16,660	1.915	8,702	.5222
38	.003965	15.72	.00004759	.00007210	21,010	1.519	13,870	.6585
39	.003531	12.47	.00003774	.00004545	26,500	1.204	22,000	.8304
40	.003145	9.888	.00002993	.00002858	33,410	.9550	34,980	1.047

* *Supplement to Transactions of American Institute of Electrical Engineers*, October, 1893.

METRIC WIRE TABLE

Calculated by the author, using the same constants and temperature co-efficients as in the Standard Copper Wire Table, p. 305.

A.W.G.	DIAME-TER	AREA	WEIGHT		LENGTH		RESISTANCE	
B. & S.	Mm.	Sq. Mm.	Kg. per M.	Kg. per Ohm	M. per Kg.	M. per Ohm	Ohms per Kg.	Ohms per M.
0000	11.7	107.2	.953	5940	1.05	6,230	.000168	.000161
000	10.4	85.0	.756	3730	1.32	4,940	.000268	.000202
00	9.27	67.4	.599	2350	1.67	3,920	.000426	.000255
0	8.25	53.5	.475	1480	2.10	3,110	.000677	.000322
1	7.35	42.4	.377	929	2.65	2,460	.00108	.000406
2	6.54	33.6	.299	584	3.35	1,950	.00171	.000512
3	5.83	26.7	.237	367	4.22	1,550	.00272	.000645
4	5.19	21.2	.188	231	5.32	1,230	.00433	.000814
5	4.62	16.8	.149	145	6.71	975	.00688	.00103
6	4.11	13.3	.118	91.4	8.46	773	.0109	.00129
7	3.67	10.6	.0938	57.5	10.7	613	.0174	.00163
8	3.26	8.37	.0744	36.2	13.5	486	.0277	.00206
9	2.91	6.63	.0590	22.7	17.0	386	.0440	.00259
10	2.59	5.26	.0468	14.3	21.4	306	.0699	.00327
11	2.31	4.17	.0371	8.99	27.0	242	.111	.00413
12	2.05	3.31	.0294	5.66	34.0	192	.177	.00520
13	1.83	2.62	.0234	3.56	42.9	153	.281	.00656
14	1.63	2.08	.0185	2.24	54.1	121	.447	.00827
15	1.45	1.65	.0147	1.41	68.2	95.9	.711	.0104
16	1.29	1.31	.0116	.885	86.0	76.0	1.13	.0132
17	1.15	1.04	.00922	.556	108	60 3	1.80	.0166
18	1.02	.823	.00732	.350	136	47.8	2.86	.0209
19	.912	.653	.00580	.220	172	37.9	4.54	.0264
20	.812	.518	.00460	.138	217	30.1	7.23	.0333
21	.723	.410	.00365	.0871	274	23.9	11.5	.0419
22	.644	.326	.00289	.0548	346	18.9	18.3	.0529
23	.573	.258	.00229	.0344	436	15 0	29.1	.0667
24	.511	.205	.00182	.0217	550	11.9	46.2	.0841
25	.455	.162	.00144	.0136	693	9.43	73.4	.106
26	.405	.129	.00114	.00856	874	7.48	117	.134
27	.361	.102	.000908	.00538	1,100	5.93	186	.169
28	.321	.081	.000720	.00339	1,390	4.70	295	.213
29	.286	.0642	.000571	.00213	1,750	3.73	470	.268
30	.255	.0510	.000453	.00134	2,210	2.96	747	.338
31	.227	.0404	.000359	.000842	2,790	2.35	1,100	.426
32	.202	.0320	.000285	.000530	3,510	1.86	1,890	.537
33	.180	.0254	.000226	.000333	4,430	1.48	3,000	.678
34	.160	.0201	.000179	.000210	5,590	1.17	4,770	.855
35	.143	.0160	.000142	.000132	7,040	.928	7,590	1.08
36	.127	.0127	.000113	.0000829	8,880	.736	12,100	1.36
37	.113	.0101	.0000893	.0000521	11,200	.584	19,200	1.71
38	.101	.00797	.0000708	.0000327	14,100	.463	30,600	2.16
39	.0897	.00632	.0000562	.0000206	17,800	.367	48,500	2 73
40	.0799	.00501	.0000445	.0000130	22,500	.291	77,100	3 44

APPROXIMATE EQUIVALENT CROSS-SECTIONS OF WIRES
(Brown and Sharpe Gauge)

Size	0000	000	00	0	1	2	3	4	5	6	7	8	9	10	11	12	13	14	15	16	17	18	19	20	21	22	23	24
0				1																								
3	4			2			1																					
4	5	4			2			1																				
5	6	5	4			2			1																			
6	8	6	5	4			2			1																		
7	10	8	6	5	4			2			1																	
8	13	10	8	6	5	4			2			1																
9	16	13	10	8	6	5	4			2			1															
10	20	16	13	10	8	6	5	4			2			1														
11	26	20	16	13	10	8	6	5	4			2			1													
12	32	26	20	16	13	10	8	6	5	4			2			1												
13	41	32	26	20	16	13	10	8	6	5	4			2			1											
14	52	41	32	26	20	16	13	10	8	6	5	4			2			1										
15	65	52	41	32	26	20	16	13	10	8	6	5	4			2			1									
16	82	65	52	41	32	26	20	16	13	10	8	6	5	4			2			1								
17	103	82	65	52	41	32	26	20	16	13	10	8	6	5	4			2			1							
18	130	103	82	65	52	41	32	26	20	16	13	10	8	6	5	4			2			1						
19	164	130	103	82	65	52	41	32	26	20	16	13	10	8	6	5	4			2			1					
20	207	164	130	103	82	65	52	41	32	26	20	16	13	10	8	6	5	4			2			1				
21	261	207	164	130	103	82	65	52	41	32	26	20	16	13	10	8	6	5	4			2			1			
22	329	261	207	164	130	103	82	65	52	41	32	26	20	16	13	10	8	6	5	4			2			1		
23	415	329	261	207	164	130	103	82	65	52	41	32	26	20	16	13	10	8	6	5	4			2			1	
24	524	415	329	261	207	164	130	103	82	65	52	41	32	26	20	16	13	10	8	6	5	4			2			1
25	661	524	415	329	261	207	164	130	103	82	65	52	41	32	26	20	16	13	10	8	6	5	4			2		
26	833	661	524	415	329	261	207	164	130	103	82	65	52	41	32	26	20	16	13	10	8	6	5	4			2	
27	1050	833	661	524	415	329	261	207	164	130	103	82	65	52	41	32	26	20	16	13	10	8	6	5	4			2
28	1324	1050	833	661	524	415	329	261	207	164	130	103	82	65	52	41	32	26	20	16	13	10	8	6	5	4		
29	1670	1324	1050	833	661	524	415	329	261	207	164	130	103	82	65	52	41	32	26	20	16	13	10	8	6	5	4	
30	2105	1670	1324	1050	833	661	524	415	329	261	207	164	130	103	82	65	52	41	32	26	20	16	13	10	8	6	5	4

BARE COPPER WIRE

B. & S. No.	d	d^2	ρ_l	ρ_w
0	.3249	.10550	.000008176	.0003071
1	.2893	.08369	.00001031	.0004883
2	.2576	.06637	.00001300	.0007765
3	.2294	.05263	.00001639	.001235
4	.2043	.04174	.00002067	.001963
5	.1819	.03310	.00002607	.003122
6	.1620	.02625	.00003287	.004963
7	.1443	.02082	.00004144	.007892
8	.1285	.01651	.00005226	.01255
9	.1144	.01309	.00006590	.01995
10	.1019	.01038	.00008310	.03173
11	.09074	.008234	.0001048	.05045
12	.08081	.006530	.0001322	.08022
13	.07196	.005178	.0001667	.1276
14	.06408	.004107	.0002101	.2028
15	.05707	.003257	.0002649	.3225
16	.05082	.002583	.0003341	.5128
17	.04526	.002048	.0004213	.8153
18	.04030	.001624	.0005312	1.296
19	.03589	.001288	.0006698	2.061
20	.03196	.001022	.000845	3.278
21	.02846	.0008101	.001065	5.212
22	.02535	.0006424	.001343	8.287
23	.02257	.0005095	.001693	13.18
24	.02010	.0004040	.002136	20.95
25	.01790	.0003204	.002693	33.32
26	.01594	.0002541	.003396	52.97
27	.01420	.0002015	.004282	84.23
28	.01264	.0001598	.005399	133.9
29	.01126	.0001267	.006808	213.0
30	.01003	.0001005	.008583	338.6
31	.008928	.00007970	.01083	538.4
32	.007950	.00006321	.01365	856.2
33	.007080	.00005013	.01722	1,361
34	.006305	.00003975	.02171	2,165
35	.005615	.00003152	.02737	3,441
36	.005000	.00002500	.03452	5,473
37	.004453	.00001983	.04352	8,702
38	.003965	.00001572	.05488	13,870
39	.003531	.00001247	.06920	22,000
40	.003145	.00000989	.08725	34,980

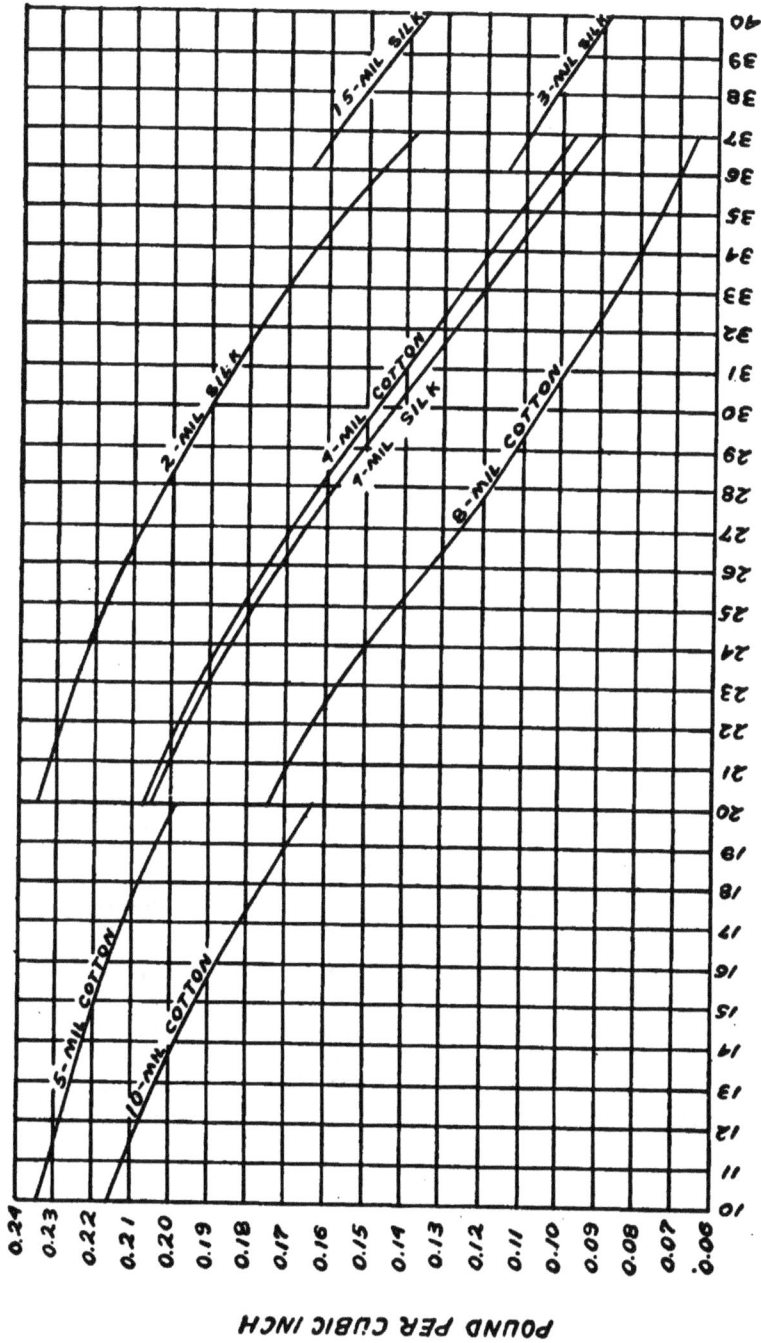

FIG. 215. — Weight per Cubic Inch (W_v) for Insulated Wires.

VALUES OF d_1 FOR DIFFERENT THICKNESSES OF INSULATION

No. B. & S.	10-MIL	5-MIL	
10	.1119	.1069	
11	.1007	.0957	
12	.0908	.0858	
13	.0820	.0770	
14	.0741	.0691	
15	.0671	.0621	
16	.0608	.0558	
17	.0553	.0503	
18	.0503	.0453	
19	.0459	.0409	
	8-MIL	4-MIL	2-MIL
20	.0400	.0360	.0340
21	.0365	.0325	.0305
22	.0334	.0294	.0274
23	.0306	.0266	.0246
24	.0281	.0241	.0221
25	.0259	.0219	.0199
26	.0239	.0200	.0179
27	.0222	.0182	.0162
28	.0206	.0166	.0146
29	.0193	.0153	.0133
30	.0180	.0140	.0120
31	.0170	.0130	.0109
32	.0160	.0120	.00995
33	.0151	.0111	.00908
34	.0143	.0103	.00830
35	.0136	.0096	.00761
36	.0130	.0090	.00700
		3-MIL	1.5-MIL
37		.00745	.0060
38		.00697	.0055
39		.00653	.00503
40		.00615	.00465

TABLE SHOWING VALUES OF N_a* (TURNS PER SQUARE INCH), FOR DIFFERENT THICKNESSES OF INSULATION

No. B. & S.	10-Mil	5-Mil	
10	80	87	
11	98	109	
12	121	135	
13	149	168	
14	182	210	
15	222	260	
16	270	321	
17	328	396	
18	396	487	
19	476	599	
	8-Mil	4-Mil	2-Mil
20	627	775	869
21	752	950	1,075
22	900	1,160	1,335
23	1,070	1,420	1,655
24	1,270	1,725	2,050
25	1,490	2,090	2,520
26	1,740	2,520	3,090
27	2,025	3,010	3,810
28	2,350	3,610	4,680
29	2,700	4,300	5,690
30	3,080	5,080	6,900
31	3,490	6,000	8,370
32	3,930	7,000	10,100
33	4,400	8,140	12,100
34	4,880	9,450	14,500
35	5,400	10,800	17,250
36	5,920	12,350	20,400
		3-Mil	1.5-Mil
37		18,000	28,200
38		20,600	33,400
39		23,450	39,500
40		26,450	46,250

* These values are for solid layer winding with no paper.

ENAMELITE WIRE *

(Approximate Values)

No. B. & S.	d_1 Total Diameter	N_a † Turns per sq. in.	ρ_v † Ohms per cu. in.	O_P Ohms per lb.
10	.1038	92.5	.00768	.0315
11	.0925	117	.0123	.0502
12	.0826	147	.0194	.0797
13	.0737	184	.0307	.1278
14	.0658	231	.0485	.2015
15	.0585	292	.0774	.3200
16	.0523	366	.1225	.509
17	.0467	458	.193	.809
18	.0418	572	.305	1.285
19	.0374	715	.479	2.042
20	.0332	907	.767	3.24
21	.0295	1,150	1.225	5.16
22	.0265	1,425	1.915	8.19
23	.0237	1,780	3.01	13.02
24	.0212	2,220	4.74	20.7
25	.0189	2,800	7.54	32.9
26	.0168	3,540	12.03	52.2
27	.0150	4,440	19.0	82.9
28	.0134	5,570	30.1	131.5
29	.0120	6,950	47.3	208.5
30	.0107	8,730	74.8	334
31	.0097	10,650	115.3	530
32	.0086	13,500	184.2	840
33	.0077	16,900	291	1,333
34	.0069	21,000	456	2,115
35	.0062	26,000	711	3,350
36	.0056	31,900	1,100	5,320
37	.0050	40,000	1,742	8,440
38	.0045	49,300	2,710	13,400

* The Acme Wire Co.
† These values are for solid layer winding with no paper.

ENAMELITE WIRE WITH 1-MIL PAPER BETWEEN LAYERS *

No. B. & S.	One Wrap		Two Wraps	
	N_a	ρ_v	N_a	ρ_v
22			1,260	1.69
23			1,545	2.61
24			1,920	4.11
25	2,550	6.9	2,420	6.5
26	3,190	10.85	3,000	10.2
27	3,900	16.7	3,660	15.6
28	4,830	26.1	4,480	24.2
29	5,900	40.2	5,430	37
30	7,225	62	6,600	56.5
31	8,800	95	8,000	86.5
32	10,800	148	9,700	133
33	13,400	230	11,900	205
34	16,500	358	14,500	316
35	20,500	560	17,600	483
36	25,000	860	21,400	740
37	29,500	1,280	25,000	1,090
38	34,000	1,860	28,400	1,560

DELTABESTON WIRE TABLE †

No. B. & S.	d	d_1	Cir. Mils	Ohms per 1000 Ft.	Feet per Lb. Bare	Feet per Lb. Deltab.	Lbs. per 1000 Ft. Deltab.
00	.3648		133,079	.0789	2.485		
0	.3250	.343	105,625	.0994	3.131		
1	.2893	.307	83,694	.1255	3.952	3.88	257.73
2	.2576	.276	66,358	.1583	4.984	4.90	204.08
3	.2294	.247	52,624	.1996	6.285	6.15	162.60
4	.2043	.220	41,738	.2516	7.924	7.70	129.87
5	.1819	.198	33,088	.3174	9.996	9.70	103.09
6	.1620	.178	26,244	.4002	12.60	12.2	81.97
7	.1443	.160	20,822	.5044	15.88	15.4	64.93
8	.1285	.142	16,512	.6361	20.03	19.5	51.28
9	.1144	.128	13,087	.8026	25.27	24.5	40.82
10	.1019	.116	10,384	1.011	31.85	30.8	32.47
11	.0907	.103	8,226	1.277	40.21	37.9	26.38
12	.0808	.093	6,527	1.609	50.66	46.7	21.41
13	.0720	.082	5,184	2.026	63.80	58.4	17.12
14	.0641	.074	4,109	2.556	80.50	73.2	13.66
15	.0571	.067	3,260	3.221	101.4	91.7	10.90
16	.0508	.061	2,581	4.070	128.2	114.4	8.74
17	.0453	.055	2,052	5.118	161.2	145	6.90
18	.0403	.050	1,624	6.466	203.7	183	5.46
19	.0359	.046	1,289	8.151	256.6	231	4.33
20	.0320	.042	1,024	10.26	323.0		

* The Acme Wire Co.   † D. & W. Fuse Co.

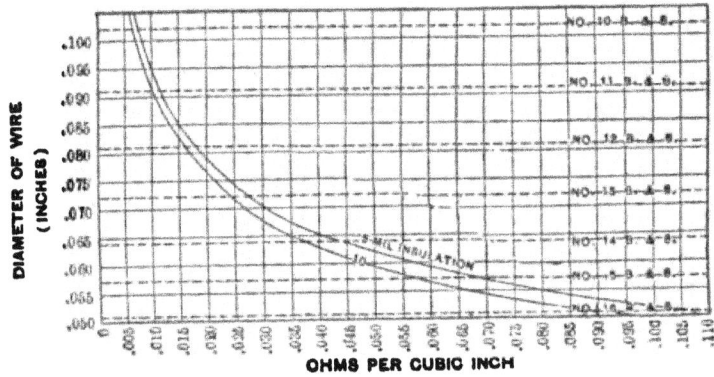

FIG. 216. — ρ_v Values. Nos. 10 to 16, B. & S.

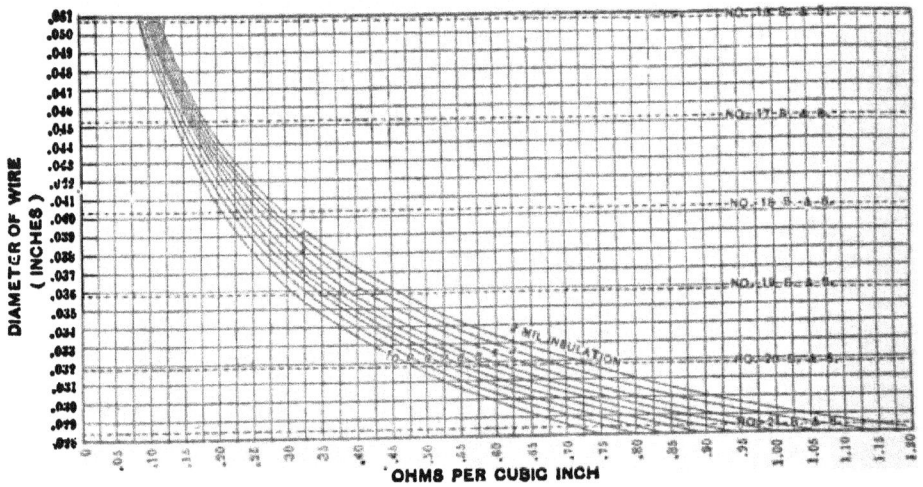

FIG. 217. — ρ_v Values. Nos. 16 to 21, B. & S.

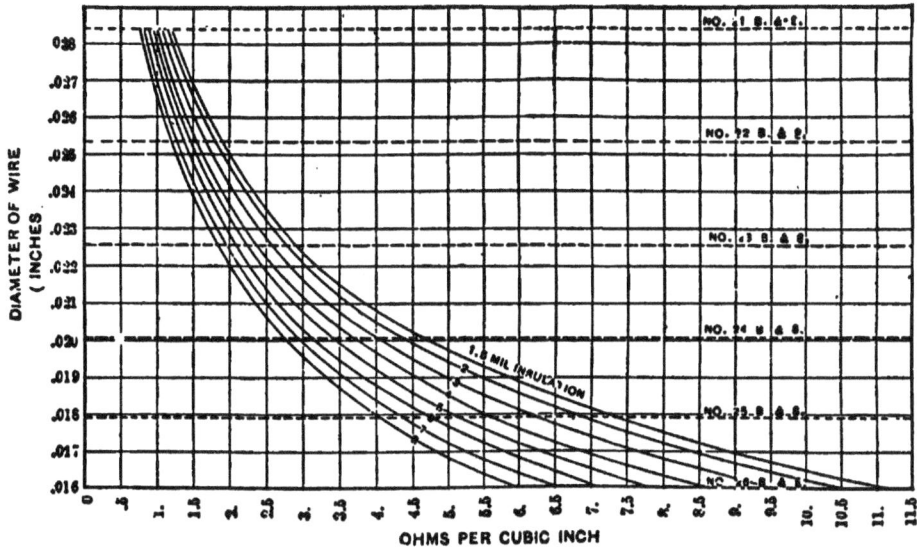

FIG. 218. — ρ_v Values. Nos. 21 to 26, B. & S.

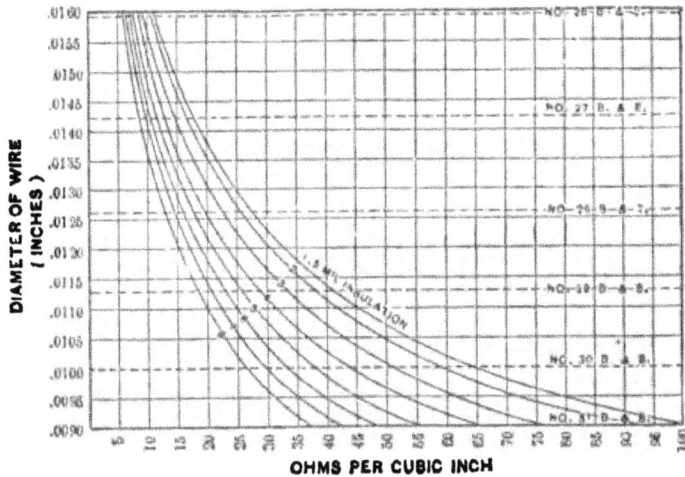

FIG. 219. — ρ_v Values. Nos. 26 to 31, B. & S.

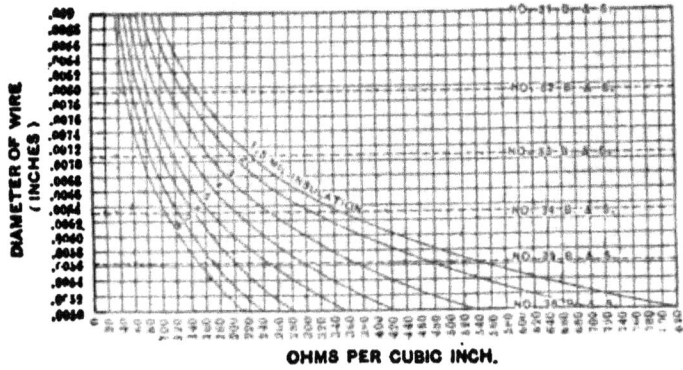

FIG. 220. — ρ_v Values.   Nos. 31 to 36, B & S.

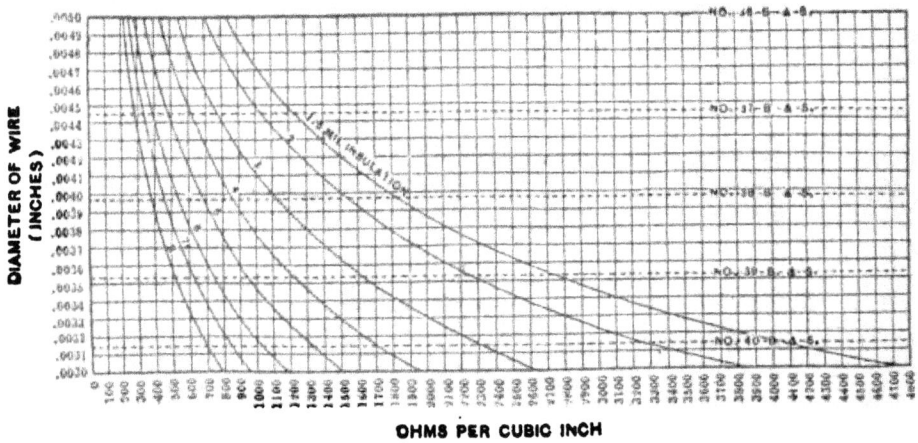

FIG. 221. — ρ_v Values.   Nos. 36 to 40, B. & S.

RESISTANCE WIRES *

(Ohms per 1000 feet)

No. B. & S.	FERRO-NICKEL	S. B.	MANGANIN
1	2.0		
2	2.5		
3	3.2		
4	4.1		
5	5.1		
6	6.5		
7	8.2		
8	10.4		
9	13.1		
10	16.3	32	
11	20.5	40	
12	25.9	51	
13	32.7	64	
14	41.5	82	
15	52.3	103	76
16	65.4	130	94
17	85	168	122
18	106	210	153
19	131	260	189
20	166	328	244
21	209	415	301
22	266	525	382
23	333	660	480
24	425	831	606
25	531	1,050	765
26	672	1,328	968
27	850	1,667	1,212
28	1,070	2,112	1,540
29	1,330	2,625	1,910
30	1,700	3,360	2,448
31	2,120	4,250	3,090
32	2,660	5,250	3,825
33	3,400	6,660	4,857
34	4,250	8,400	6,166
35	5,400	10,700	7,796
36	6,800	13,440	9,792
37		16,640	12,363
38		21,000	15,692
39		27,540	19,742
40		37,300	24,980

* Driver-Harris Wire Co.

PROPERTIES OF "NICHROME" RESISTANCE WIRE *

Resistance per mil-foot at 75° Fahr. — 575 ohms.
Temperature coefficient — .00024 per degree Fahr.
Specific gravity — 8.15.

No. B. & S.	Diameter in Inches	Area in Circular Mils C. M.-D²	Resistance per 1000 ft. at 75° F.	Weights per 1000 ft. Bare	Ohms per Pound
1	.289	83,521	6.9	231	.029
2	.258	66,564	8.5	184	.046
3	.229	52,441	11.0	145	.076
4	.204	41,616	13.8	115	.12
5	.182	33,124	17.3	92	.188
6	.162	26,244	21.9	73	.300
7	.144	20,736	27.7	57	.485
8	.128	16,384	35.1	45	.78
9	.114	12,996	44.2	36	1.22
10	.102	10,404	55.2	29	1.90
11	.091	8,281	69.5	23.	3.02
12	.081	6,561	87.7	18.	4.85
13	.072	5,184	110	14.3	7.70
14	.064	4,096	140	11.3	12.4
15	.057	3,249	177	9.2	19.3
16	.051	2,601	220	7.2	30.6
17	.045	2,025	284	5.6	50.7
18	.040	1,600	359	4.42	81.3
19	.036	1,296	443	3.58	124
20	.032	1,024	560	2.83	198
21	.0285	812.3	710	2.24	317
22	.0253	640.1	900	1.77	508
23	.0226	510.8	1,125	1.41	800
24	.0201	404	1,420	1.12	1,270
25	.0179	320.4	1,795	.89	2,020
26	.0159	252.8	2,275	.70	3,250
27	.0142	201.6	2,850	.56	5,100
28	.0126	158.8	3,620	.44	8,200
29	.0113	127.7	4,500	.35	12,850
30	.010	100	5,750	.276	20,800
31	.0089	79.2	7,270	.219	33,200
32	.008	64	9,000	.177	50,800
33	.0071	50.4	11,400	.139	82,000
34	.0063	39.7	14,500	.11	132,000
35	.0056	31.4	18,300	.087	210,000
36	.005	25	23,000	.069	333,000
37	.0045	20.2	28,500	.056	508,000
38	.004	16	36,000	.045	800,000
39	.0035	12.2	47,000	.034	1,383,000
40	.003	9	64,000	.025	2,560,000

* Driver-Harris Wire Co.

PROPERTIES OF "CLIMAX" RESISTANCE WIRE*

Resistance per mil-foot at 75° Fahr. — 525 ohms.
Temperature coefficient — .0003 per degree Fahr.
Specific gravity — 8.137.

No. B. & S.	Diameter in Inches	Area in Circular Mils C. M.-D²	Resistance per 1000 ft. at 75° F.	Weights per 1000 ft. Bare	Ohms per Pound
1	.289	83,521	6.2	231	.026
2	.258	66,564	7.9	184	.041
3	.229	52,441	10.0	145	.066
4	.204	41,616	12.6	115	.105
5	.182	33,124	15.8	92	.165
6	.162	26,244	20.0	73	.263
7	.144	20,736	25.3	57	.427
8	.128	16,384	32.	45	.685
9	.114	12,996	40.4	36	1.08
10	.102	20,404	50.4	29	1.65
11	.091	8,281	63.4	23	2.70
12	.081	6,561	80	18	4.27
13	.072	5,184	101	14.3	6.85
14	.064	4,096	128	11.3	10.9
15	.057	3,249	161	9.2	16.9
16	.051	2,601	202	7.2	27.0
17	.045	2,025	258	5.6	44.5
18	.040	1,600	328	4.42	71.3
19	.036	1,296	404	3.58	108
20	.032	1,024	510	2.83	174
21	.0285	812.3	646	2.24	284
22	.0253	640.1	820	1.77	456
23	.0226	510.8	1,027	1.41	720
24	.0201	404	1,290	1.12	1,142
25	.0179	320.4	1,640	.89	1,810
26	.0159	252.8	2,080	.70	2,920
27	.0142	201.6	2,580	.56	4,570
28	.0126	158.8	3,300	.44	7,400
29	.0113	127.7	4,100	.35	11,560
30	.010	100	5,250	.276	18,785
31	.0089	79.2	6,620	.219	29,800
32	.008	64	8,200	.177	45,265
33	.0071	50.4	10,410	.139	73,214
34	.0063	39.7	13,220	.11	118,300
35	.0056	31.4	16,720	.087	189,000
36	.005	25	21,000	.069	300,000
37	.0045	20.2	26,000	.056	468,000
38	.004	16	33,000	.045	733,000
39	.0035	12.2	43,000	.034	1,264,000
40	.003	9	58,000	.025	2,320,000

* Driver-Harris Wire Co.

PROPERTIES OF "ADVANCE" RESISTANCE WIRE*

Resistance per mil-foot at 75° Fahr. — 294 ohms.
Temperature coefficient — Nil. Specific gravity — 8.9.

B. & S. Gauge No.	Diameter in Inches	Area in Circular Mils C. M.-D²	Resistance per 1000 ft. at 75° F.	Weights per 1000 ft. Bare	Ohms per Pound
1	.289	83,521	3.52	253	.01365
2	.258	66,564	4.42	201	.02174
3	.229	52,441	5.61	159	.03458
4	.204	41,616	7.07	126	.05496
5	.182	33,124	8.88	100	.08742
6	.162	26,244	11.21	79	.13896
7	.144	20,736	14.19	63	.2209
8	.128	16,384	17.9	50	.3514
9	.114	12,996	22.6	39	.5586
10	.102	10,404	28.	32	.888
11	.091	8,281	35.5	25	1.412
12	.081	6,561	44.8	20	2.246
13	.072	5,184	56.7	15.7	3.573
14	.064	4,096	71.7	12.4	5.678
15	.057	3,249	90.4	9.8	9.03
16	.051	2,601	113	7.8	14.358
17	.045	2,025	145	6.2	22.828
18	.040	1,600	184	4.9	36.288
19	.036	1,296	226	3.9	57.708
20	.032	1,024	287	3.1	91.784
21	.0285	812.3	362	2.5	145.93
22	.0253	640.1	460	1.9	232.03
23	.0226	510.8	575	1.5	369.04
24	.0201	404	725	1.2	586.6
25	.0179	320.4	919	.97	932.96
26	.0159	252.8	1,162	.77	1,483.16
27	.0142	201.6	1,455	.61	2,358
28	.0126	158.8	1,850	.48	3,749
29	.0113	127.7	2,300	.38	5,964
30	.010	100	2,940	.30	9,470
31	.0089	79.2	3,680	.24	15,075
32	.008	64	4,600	.19	23,973
33	.0071	50.4	5,830	.15	38,108
34	.0063	39.7	7,400	.12	60,620
35	.0056	31.4	9,360	.095	96,340
36	.005	25	11,760	.076	153,240
37	.0045	20.2	14,550	.060	243,650
38	.004	16	18,375	.047	388,360
39	.0035	12.2	24,100	.038	616,000
40	.003	9	32,660	.028	1,183,000

* Driver-Harris Wire Co.

"MONEL" WIRE*

Resistance per mil-foot — 256 ohms.
Temperature coefficient — .0011.   Specific gravity — 8.9.

No. B. & S.	Diameter in Inches	Area in Circular Mils C. M.-D²	Resistance per 1000 ft. at 75° F.	Weights per 1000 ft. Bare	Ohms per Pound
0	.325	105,625	2.4	317	.0075
1	.289	83,521	3.0	253	.0118
2	.258	66,564	3.8	201	.0189
3	.229	52,441	4.8	159	.0301
4	.204	41,616	6.1	126	.0484
5	.182	33,124	7.7	100	.077
6	.162	26,244	9.8	79	.124
7	.144	20,736	12.3	63	.196
8	.128	16,384	15.6	50	.312
9	.114	12,996	19.7	39	.505
10	.102	10,404	24.6	32	.769
11	.091	8,281	30.9	25	1.235
12	.081	6,561	39.1	20	1.955
13	.072	5,184	49.4	15.8	3.142
14	.064	4,096	62.6	12.4	5.05
15	.057	3,249	78.9	9.8	8.04
16	.051	2,601	98.6	7.8	12.62
17	.045	2,025	121	6.2	19.51
18	.040	1,600	160	4.9	32.69
19	.036	1,296	198	3.9	50.77
20	.032	1,024	250	3.1	80.64
21	.0285	812.3	315	2.5	126
22	.0253	640.1	400	1.9	210.52
23	.0226	510.8	502	1.5	334.66
24	.0201	404	635	1.2	529.16
25	.0179	320.4	800	.97	824.7
26	.0159	252.8	991	.77	1,287
27	.0142	201.6	1,272	.61	2,085
28	.0126	158.8	1,615	.48	3,365
29	.0113	127.7	2,009	.38	5,286
30	.010	100	2,566	.30	8,543
31	.0089	79.2	3,239	.24	13,495
32	.008	64	4,009	.19	21,000
33	.0071	50.4	5,091	.15	33,940
34	.0063	39.7	6,463	.12	53,858
35	.0056	31.4	8,172	.095	86,021
36	.005	25	10,260	.076	135,000
37	.0045	20.2	12,700	.060	267,166
38	.004	16	16,030	.047	341,063
39	.0035	12.2	21,030	.038	553,421
40	.003	9	28,510	.028	1,018,214

* Driver-Harris Wire Co.

TABLE SHOWING THE DIFFERENCE BETWEEN WIRE GAUGES

No.	Brown & Sharpe's	London	Birmingham or Stubs	W. & M. and Roebling	New British Standard
0000	.460	.454	.454	.393	.400
000	.40964	.425	.425	.362	.372
00	.36480	.380	.380	.331	.348
0	.32495	.340	.340	.307	.324
1	.28930	.300	.300	.283	.300
2	.25763	.284	.284	.263	.276
3	.22942	.259	.259	.244	.252
4	.20431	.238	.238	.225	.232
5	.18194	.220	.220	.207	.212
6	.16202	.203	.203	.192	.192
7	.14428	.180	.180	.177	.176
8	.12849	.165	.165	.162	.160
9	.11443	.148	.148	.148	.144
10	.10189	.134	.134	.135	.128
11	.09074	.120	.120	.120	.116
12	.08081	.109	.109	.105	.104
13	.07199	.095	.095	.092	.092
14	.06408	.083	.083	.080	.080
15	.05706	.072	.072	.072	.072
16	.05082	.065	.065	.063	.064
17	.04525	.058	.058	.054	.056
18	.04030	.049	.049	.047	.048
19	.03589	.040	.042	.041	.040
20	.03196	.035	.035	.035	.036
21	.02846	.0315	.032	.032	.032
22	.025347	.0295	.028	.028	.028
23	.022571	.027	.025	.025	.024
24	.0201	.025	.022	.023	.022
25	.0179	.023	.020	.020	.020
26	.01594	.0205	.018	.018	.018
27	.014195	.01875	.016	.017	.0164
28	.012641	.0165	.014	.016	.0148
29	.011257	.0155	.013	.015	.0136
30	.010025	.01375	.012	.014	.0124
31	.008928	.01225	.010	.0135	.0116
32	.00795	.01125	.009	.013	.0108
33	.00708	.01025	.008	.011	.010
34	.0063	.0095	.007	.010	.0092
35	.00561	.009	.005	.0095	.0084
36	.005	.0075	.004	.009	.0076
37	.00445	.0065		.0085	.0068
38	.003965	.00575		.008	.006
39	.003531	.005		.0075	.0052
40	.003144	.0045		.007	.0048

PERMEABILITY TABLE *

DENSITY OF MAGNETIZATION		PERMEABILITY, μ			
$\mathcal{B}''$ Lines per Square Inch	$\mathcal{B}$ Lines per Square Centimeter	Annealed Wrought Iron	Commercial Wrought Iron	Gray Cast Iron	Ordinary Cast Iron
20,000	3,100	2,600	1,800	850	650
25,000	3,875	2,900	2,000	800	700
30,000	4,650	3,000	2,100	600	770
35,000	5,425	2,950	2,150	400	800
40,000	6,200	2,900	2,130	250	770
45,000	6,975	2,800	2,100	140	730
50,000	7,750	2,650	2,050	110	700
55,000	8,525	2,500	1,980	90	600
60,000	9,300	2,300	1,850	70	500
65,000	10,100	2,100	1,700	50	450
70,000	10,850	1,800	1,550	35	350
75,000	11,650	1,500	1,400	25	250
80,000	12,400	1,200	1,250	20	200
85,000	13,200	1,000	1,100	15	150
90,000	14,000	800	900	12	100
95,000	14,750	530	680	10	70
100,000	15,500	360	500	9	50
105,000	16,300	260	360		
110,000	17,400	180	260		
115,000	17,800	120	190		
120,000	18,600	80	150		
125,000	19,400	50	120		
130,000	20,150	30	100		
135,000	20,900	20	85		
140,000	21,700	15	75		

* Wiener, Dynamo Electric Machines.

TRACTION TABLE

$\mathcal{B}$ LINES PER SQUARE CENTIMETER	TRACTION IN KILOGRAMS PER SQUARE CENTIMETER	$\mathcal{B}''$ LINES PER SQUARE INCH	TRACTION IN POUNDS PER SQUARE INCH
1,000	.0406	10,000	1.386
2,000	.1622	15,000	3.119
3,000	.3650	20,000	5.545
4,000	.6490	25,000	8.664
5,000	1.014	30,000	12.48
6,000	1.460	35,000	16.98
7,000	1.987	40,000	22.18
8,000	2.596	45,000	28.07
9,000	3.285	50,000	34.66
10,000	4.056	55,000	41.93
11,000	4.908	60,000	49.91
12,000	5.841	65,000	58.57
13,000	6.855	70,000	67.93
14,000	7.950	75,000	77.99
15,000	9.126	80,000	88.72
16,000	10.38	85,000	100.1
17,000	11.72	90,000	112.3
18,000	13.14	95,000	125.1
19,000	14.64	100,000	138.6
*20,000	16.22	105,000	152.8
21,000	17.89	110,000	167.8
22,000	19.63	115,000	183.3
23,000	21.46	120,000	199.6
24,000	23.36	125,000	210.6
25,000	25.35	130,000	234.3

* The limit of magnetization for wrought iron is 20,200 lines per square centimeter, or 130,000 per square inch. These values are approximate.

INSULATING MATERIALS *

(Uniform thickness and insulation)

	MATERIAL	GRADE	THICKNESS IN MILS	PUNCTURE TEST IN VOLTS
Furnished in rolls 36″ wide. Any length desired.	Linen	A	6–7	6,000
	Linen	AA	5	5,000
	Linen	B	9–10	9,000
	Linen	C	11–12	12,000
	Canvas	A	10–11	8,000
	Canvas	B	15–16	10,000
	Black insulating cloth		7, 10, and 12	1,500 per mil of thickness
Furnished in sheets 36″ × 36″.	Paper	A	5–6	5,000
	Paper	B	7–8	8,000
	Paper	C	10–11	12,000
	Red Rope Paper	A	7–8	7,000
	Bond Paper (also 17″ × 20″)	A	4–5	4,000

* Pittsburg Insulating Company.

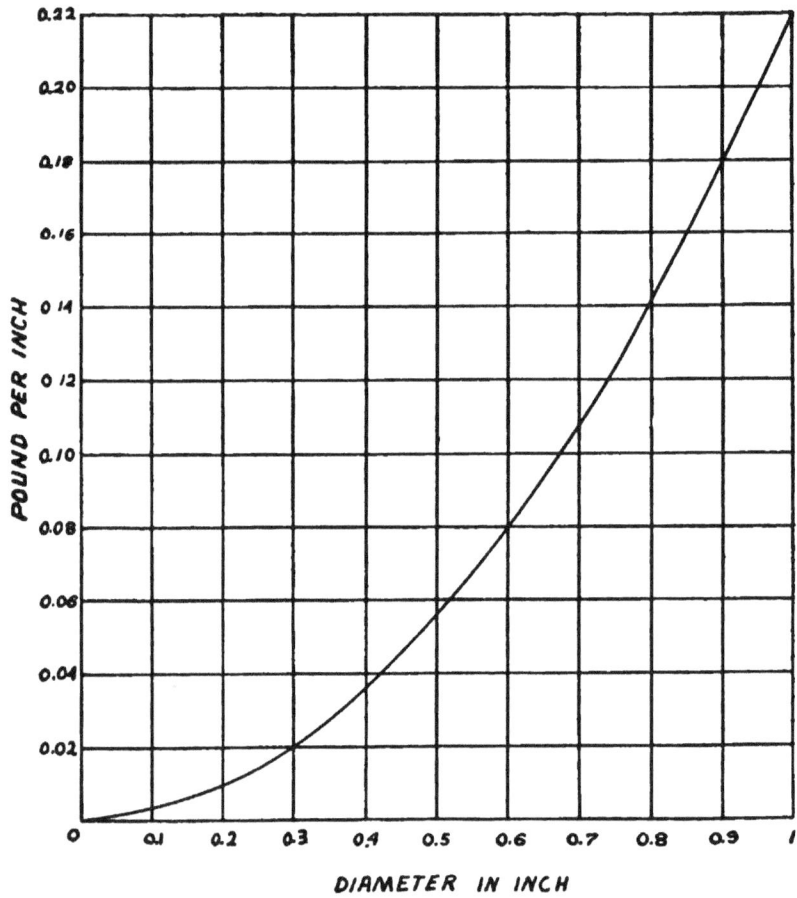

FIG. 222. — Weight per Unit Length of Plunger.

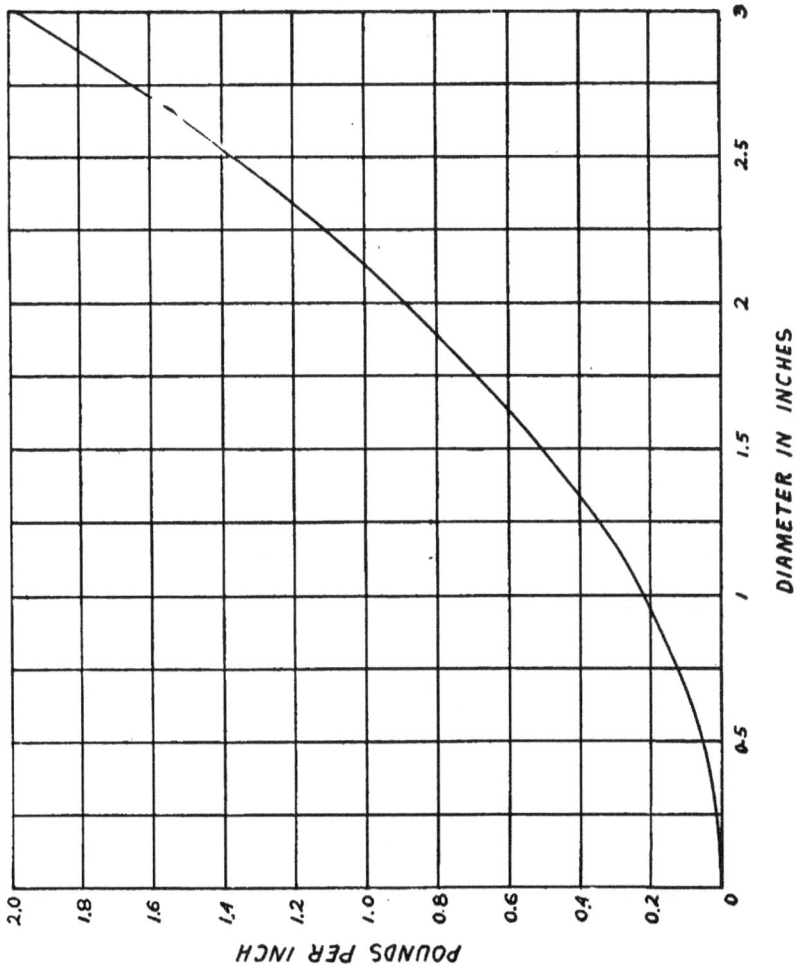

FIG. 223. — Weight per Unit Length of Plunger.

INSIDE AND OUTSIDE DIAMETERS OF BRASS TUBING
(Inches)

No. 10 (B. & S.) Wall		No. 12 (B. & S.) Wall	
Outside	Inside	Outside	Inside
$1\frac{1}{4}$	1.05	$\frac{5}{8}$	.465
$1\frac{3}{8}$	1.175	$\frac{3}{4}$	.59
$1\frac{5}{8}$	1.425	$\frac{7}{8}$	.715
		1	.84
		$1\frac{1}{8}$	.965

No. 18 (B. & S.) Wall		No. 24 (B. & S) Wall	
Outside	Inside	Outside	Inside
$\frac{1}{2}$	.42	$\frac{1}{2}$	.46
$\frac{5}{8}$	.545	$\frac{5}{8}$	.585
$\frac{3}{4}$	.67	$\frac{3}{4}$	.71
$\frac{7}{8}$	.795	$\frac{7}{8}$	.835
1	.92		
$1\frac{1}{8}$	1.045		
$1\frac{1}{4}$	1.17		
$1\frac{1}{2}$	1.42		
$1\frac{3}{4}$	1.67		
2	1.92		

DECIMAL EQUIVALENTS

8ths	16ths	32ds	64ths	Decimal Equivalent	8ths	16ths	32ds	64ths	Decimal Equivalent
			1..	.015625				33..	.515625
		1..		.03125			17..		.53125
			3..	.046875				35..	.546875
	1..			.0625		9..			.5625
			5..	.078125				37..	.578125
		3..		.09375			19..		.59375
			7..	.109375				39..	.609375
1..				.125	5..				.625
			9..	.140625				41..	.640625
		5..		.15625			21..		.65625
			11..	.171875				43..	.671875
	3..			.1875		11..			.6875
			13..	.203125				45..	.703125
		7..		.21875			23..		.71875
			15..	.234375				47..	.734375
				.25					.75
			17..	.265625				49..	.765625
		9..		.28125			25..		.78125
			19..	.296875				51..	.796875
	5..			.3125		13..			.8125
			21..	.328125				53..	.828125
		11..		.34375			27..		.84375
			23..	.359375				55..	.859375
3..				.375	7..				.875
			25..	.390625				57..	.890625
		13..		.40625			29..		.90625
			27..	.421875				59..	.921875
	7..			.4375		15..			.9375
			29..	.453125				61..	.953125
		15..		.46875			31..		.96875
			31..	.484375				63..	.984375
				.500000					

LOGARITHMS

No.	0	1	2	3	4	5	6	7	8	9
10	0000	0043	0086	0128	0170	0212	0253	0294	0334	0374
11	0414	0453	0492	0531	0569	0607	0645	0682	0719	0755
12	0792	0828	0864	0899	0934	0969	1004	1038	1072	1106
13	1139	1173	1206	1239	1271	1303	1335	1367	1399	1430
14	1461	1492	1523	1553	1584	1614	1644	1673	1703	1732
15	1761	1790	1818	1847	1875	1903	1931	1959	1987	2014
16	2041	2068	2095	2122	2148	2175	2201	2227	2253	2279
17	2304	2330	2355	2380	2405	2430	2455	2480	2504	2529
18	2553	2577	2601	2625	2648	2672	2695	2718	2742	2765
19	2788	2810	2833	2856	2878	2900	2923	2945	2967	2989
20	3010	3032	3054	3075	3096	3118	3139	3160	3181	3201
21	3222	3243	3263	3284	3304	3324	3345	3365	3385	3404
22	3424	3444	3464	3483	3502	3522	3541	3560	3579	3598
23	3617	3636	3655	3674	3692	3711	3729	3747	3766	3784
24	3802	3820	3838	3856	3874	3892	3909	3927	3945	3962
25	3979	3997	4014	4031	4048	4065	4082	4099	4116	4133
26	4150	4166	4183	4200	4216	4232	4249	4265	4281	4298
27	4314	4330	4346	4362	4378	4393	4409	4425	4440	4456
28	4472	4487	4502	4518	4533	4548	4564	4579	4594	4609
29	4624	4639	4654	4669	4683	4698	4713	4728	4742	4757
30	4771	4786	4800	4814	4829	4843	4857	4871	4886	4900
31	4914	4928	4942	4955	4969	4983	4997	5011	5024	5038
32	5051	5065	5079	5092	5105	5119	5132	5145	5159	5172
33	5185	5198	5211	5224	5237	5250	5263	5276	5289	5302
34	5315	5328	5340	5353	5366	5378	5391	5403	5416	5428
35	5441	5453	5465	5478	5490	5502	5514	5527	5539	5551
36	5563	5575	5587	5599	5611	5623	5635	5647	5658	5670
37	5682	5694	5705	5717	5729	5740	5752	5763	5775	5786
38	5798	5809	5821	5832	5843	5855	5866	5877	5888	5899
39	5911	5922	5933	5944	5955	5966	5977	5988	5999	6010
40	6021	6031	6042	6053	6064	6075	6085	6096	6107	6117
41	6128	6138	6149	6160	6170	6180	6191	6201	6212	6222
42	6232	6243	6253	6263	6274	6284	6294	6304	6314	6325
43	6335	6345	6355	6365	6375	6385	6395	6405	6415	6425
44	6435	6444	6454	6464	6474	6484	6493	6503	6513	6522
45	6532	6542	6551	6561	6571	6580	6590	6599	6609	6618
46	6628	6637	6646	6656	6665	6675	6684	6693	6702	6712
47	6721	6730	6739	6749	6758	6767	6776	6785	6794	6803
48	6812	6821	6830	6839	6848	6857	6866	6875	6884	6893
49	6902	6911	6920	6928	6937	6946	6955	6964	6972	6981
50	6990	6998	7007	7016	7024	7033	7042	7050	7059	7067
51	7076	7084	7093	7101	7110	7118	7126	7135	7143	7152
52	7160	7168	7177	7185	7193	7202	7210	7218	7226	7235
53	7243	7251	7259	7267	7275	7284	7292	7300	7308	7316
54	7324	7332	7340	7348	7356	7364	7372	7380	7388	7396
No.	0	1	2	3	4	5	6	7	8	9

LOGARITHMS

No.	0	1	2	3	4	5	6	7	8	9
55	7404	7412	7419	7427	7435	7443	7451	7459	7466	7474
56	7482	7490	7497	7505	7513	7520	7528	7536	7543	7551
57	7559	7566	7574	7582	7589	7597	7604	7612	7619	7627
58	7634	7642	7649	7657	7664	7672	7679	7686	7694	7701
59	7709	7716	7723	7731	7738	7745	7752	7760	7767	7774
60	7782	7789	7796	7803	7810	7818	7825	7832	7839	7846
61	7853	7860	7868	7875	7882	7889	7896	7903	7910	7917
62	7924	7931	7938	7945	7952	7959	7966	7973	7980	7987
63	7993	8000	8007	8014	8021	8028	8035	8041	8048	8055
64	8062	8069	8075	8082	8089	8096	8102	8109	8116	8122
65	8129	8136	8142	8149	8156	8162	8169	8176	8182	8189
66	8195	8202	8209	8215	8222	8228	8235	8241	8248	8254
67	8261	8267	8274	8280	8287	8293	8299	8306	8312	8319
68	8325	8331	8338	8344	8351	8357	8363	8370	8376	8382
69	8388	8395	8401	8407	8414	8420	8426	8432	8439	8445
70	8451	8457	8463	8470	8476	8482	8488	8494	8500	8506
71	8513	8519	8525	8531	8537	8543	8549	8555	8561	8567
72	8573	8579	8585	8591	8597	8603	8609	8615	8621	8627
73	8633	8639	8645	8651	8657	8663	8669	8675	8681	8686
74	8692	8698	8704	8710	8716	8722	8727	8733	8739	8745
75	8751	8756	8762	8768	8774	8779	8785	8791	8797	8802
76	8808	8814	8820	8825	8831	8837	8842	8848	8854	8859
77	8865	8871	8876	8882	8887	8893	8899	8904	8910	8915
78	8921	8927	8932	8938	8943	8949	8954	8960	8965	8971
79	8976	8982	8987	8993	8998	9004	9009	9015	9020	9025
80	9031	9036	9042	9047	9053	9058	9063	9069	9074	9079
81	9085	9090	9096	9101	9106	9112	9117	9122	9128	9133
82	9138	9143	9149	9154	9159	9165	9170	9175	9180	9186
83	9191	9196	9201	9206	9212	9217	9222	9227	9232	9238
84	9243	9248	9253	9258	9263	9269	9274	9279	9284	9289
85	9294	9299	9304	9309	9315	9320	9325	9330	9335	9340
86	9345	9350	9355	9360	9365	9370	9375	9380	9385	9390
87	9395	9400	9405	9410	9415	9420	9425	9430	9435	9440
88	9445	9450	9455	9460	9465	9469	9474	9479	9484	9489
89	9494	9499	9504	9509	9513	9518	9523	9528	9533	9538
90	9542	9547	9552	9557	9562	9566	9571	9576	9581	9586
91	9590	9595	9600	9605	9609	9614	9619	9624	9628	9633
92	9638	9643	9647	9652	9657	9661	9666	9671	9675	9680
93	9685	9689	9694	9699	9703	9708	9713	9717	9722	9727
94	9731	9736	9741	9745	9750	9754	9759	9763	9768	9773
95	9777	9782	9786	9791	9795	9800	9805	9809	9814	9818
96	9823	9827	9832	9836	9841	9845	9850	9854	9859	9863
97	9868	9872	9877	9881	9886	9890	9894	9899	9903	9908
98	9912	9917	9921	9926	9930	9934	9939	9943	9948	9952
99	9956	9961	9965	9969	9974	9978	9983	9987	9991	9996
No.	0	1	2	3	4	5	6	7	8	9

COMPARISON OF MAGNETIC AND ELECTRIC CIRCUIT RELATIONS

MAGNETIC	ELECTRIC
Magnetomotive force ($\mathcal{F}$)	Electromotive force (E)
Reluctance ($\mathcal{R}$)	Resistance (R)
Flux (ϕ)	Current (I)
$$\phi = \frac{\mathcal{F}}{\mathcal{R}}$$	$$I = \frac{E}{R}$$
Intensity of field or magnetizing force ($\mathcal{H}$)	Difference of potential (E_1)
$$\mathcal{H} = \frac{\mathcal{F}}{l_c}$$	$$E_1 = \frac{E}{l_w}$$
Induction or flux density ($\mathcal{B}$)	Current density (I_d)
$$\mathcal{B} = \frac{\phi}{A}$$	$$I_d = \frac{I}{A}$$
Reluctivity, specific reluctance (v)	Resistivity, specific resistance (ρ)
$$v = \frac{A\mathcal{R}}{l_c}$$	$$\rho = \frac{AR}{l_w}$$
Permeance, reciprocal of reluctance ($\mathcal{P}$)	Conductance, reciprocal of resistance (G)
$$\mathcal{P} = \frac{1}{\mathcal{R}}$$	$$G = \frac{1}{R}$$
Permeability, specific permeance, reciprocal of reluctivity (μ)	Conductivity, specific conductance, reciprocal of resistivity (γ)
$$\mu = \frac{1}{v} = \frac{l_c}{A\mathcal{R}}$$	$$\gamma = \frac{1}{\rho} = \frac{l_w}{AR}$$

14. TRIGONOMETRIC FUNCTIONS.

ANG. DEG.	SINE.	TAN.	COTAN.	COS.	ANG.
0	0.0000	0.0000	0.0000	1.0000	90
1	0.0175	0.0175	57.2400	0.9998	89
2	0.0349	0.0349	28.6363	0.9994	88
3	0.0523	0.0524	19.0811	0.9986	87
4	0.0698	0.0699	14.3007	0.9976	86
5	0.0872	0.0875	11.4301	0.9962	85
6	0.1045	0.1051	9.5144	0.9945	84
7	0.1219	0.1228	8.1443	0.9925	83
8	0.1392	0.1405	7.1154	0.9903	82
9	0.1564	0.1584	6.3138	0.9877	81
10	0.1736	0.1763	5.6713	0.9848	80
11	0.1908	0.1944	5.1446	0.9816	79
12	0.2079	0.2126	4.7046	0.9781	78
13	0.2250	0.2309	4.3315	0.9744	77
14	0.2419	0.2493	4.0108	0.9703	76
15	0.2588	0.2679	3.7321	0.9659	75
16	0.2756	0.2867	3.4874	0.9613	74
17	0.2924	0.3057	3.2709	0.9563	73
18	0.3090	0.3249	3.0777	0.9511	72
19	0.3256	0.3443	2.9042	0.9445	71
20	0.3420	0.3640	2.7475	0.9397	70
21	0.3584	0.3839	2.6051	0.9336	69
22	0.3746	0.4040	2.4751	0.9272	68
ANG.	COSINE.	COTAN.	TAN.	SINE.	ANG. DEG.

ANG. DEG.	SINE.	TAN.	COTAN.	COS.	ANG.
23	0.3907	0.4245	2.3559	0.9205	67
24	0.4067	0.4452	2.2460	0.9135	66
25	0.4226	0.4663	2.1445	0.9063	65
26	0.4384	0.4877	2.0503	0.8988	64
27	0.4540	0.5095	1.9626	0.8910	63
28	0.4695	0.5317	1.8807	0.8829	62
29	0.4848	0.5543	1.8040	0.8746	61
30	0.5000	0.5774	1.7321	0.8660	60
31	0.5150	0.6009	1.6643	0.8572	59
32	0.5299	0.6249	1.6003	0.8480	58
33	0.5446	0.6494	1.5399	0.8387	57
34	0.5592	0.6745	1.4826	0.8290	56
35	0.5736	0.7002	1.4281	0.8192	55
36	0.5878	0.7265	1.3764	0.8090	54
37	0.6018	0.7536	1.3270	0.7986	53
38	0.6157	0.7813	1.2799	0.7880	52
39	0.6293	0.8098	1.2349	0.7771	51
40	0.6428	0.8391	1.1918	0.7660	50
41	0.6561	0.8693	1.1504	0.7547	49
42	0.6691	0.9004	1.1106	0.7431	48
43	0.6820	0.9325	1.0724	0.7314	47
44	0.6947	0.9657	1.0355	0.7193	46
45	0.7071	1.0000	1.0000	0.7071	45
ANG.	COSINE.	COTAN.	TAN.	SINE.	ANG. DEG.

INDEX

A

Absolute units, 2.
Action, rapid, slow, 184.
Active pressure, 157.
Activity, 237–245.
Advance wire, 218.
Aging of magnets, 12.
Air-core, 76–78.
Air-gap, 10, 33, 113, 119.
Air, permeability of, 104.
Alternating-current (A. C.),
  electromagnet calculations, 175.
  electromagnets, 168.
  horseshoe electromagnet, 174.
  iron-clad solenoid, 171.
  plunger electromagnet, 172.
  solenoid, 168.
  windings, 181.
Alternating currents, 154.
  e. m. f., 155.
Alternation, 155.
American wire gauge, 215.
Ampere, 17.
Ampere-turn, 27.
Ampere-turns, 68, 79, 142, 239, 246, 248.
  per centimeter length, 34, 68.
Angles, 121, 123.
Angular velocity, 159.
Annealing, 192.
Area,
  of plunger, 77, 79, 83, 108.
  polar, 114, 120.
Armature, 14, 132, 133.
Armatures, external, 131.
Artificial magnet, 8, 10.
Asbestos, 221.
Attraction, 25, 41, 76, 142.
  mutual, 41.
Average radius, 65.

B

Back iron, 133.
Baker, H. S., 32.
Bar electromagnet, 132.
Bar permanent magnet, 13, 64.
Billet magnet, 137.
Bobbins, 195.
British thermal unit, 296.
Bulging of lines, 113.

C

Calorie, 296.
Capacity, 160, 162.
Carichoff, E. R., 113, 120.
Carrying capacity of wires, 219.
Cast iron, 191.
Cast steel, 191.
Centigrade scale, 297.
Centimeter, 2, 3, 15.
C. G. S. system, 2.
Circles of force, 25.
Circuit,
  electric, 17, 27, 153.
  magnetic, 9, 27, 34, 76, 153.
  return, 15, 102, 111.
  shunt, 20.
Circular winding, 257.
Climax wire, 218.
Closed-ring magnet, 9.
Coefficient,
  leakage, 37.
  of self-induction, 149.
  temperature, 19, 210.
Coercive force, 12.
Coil-and-plunger, 41.
Coil, coils,
  dimensions of, 55, 56, 71, 108.
  exciting, 126.

www.ingramcontent.com/pod-product-compliance
Lightning Source LLC
Chambersburg PA
CBHW051203200326
41519CB00025B/6993

9 7 8 1 6 0 3 8 6 1 6 1 8